THE HISTORY OF HEREDITARIAN THOUGHT

A thirty-two volume reprint
series presenting some of the
classic books in this
intellectual tradition

Edited by Charles Rosenberg
The University of Pennsylvania

A GARLAND SERIES

MARRIAGE
AND
DISEASE

S.A.K. Strahan

GARLAND PUBLISHING, INC.
NEW YORK • LONDON
1984

For a complete list of the titles in this series see the final
pages of this volume.

This facsimile has been made from a copy in the
Yale Medical School Library.

Library of Congress Cataloging in Publication Data

Strahan, S. A. K. (Samuel Alexander Kenny), d. 1902.
  Marriage and disease.

  (The History of hereditarian thought)
  Reprint. Originally published: New York : Appleton,
1892.
  Includes index.
  1. Medical genetics.   2. Marriage.   I. Title.
II. Series.
RB155.S9   1984   616'.042   83-48559
ISBN 0-8240-5829-1 (alk. paper)

The volumes in this series are printed on acid-free,
250-year-life paper.

Printed in the United States of America

MARRIAGE AND DISEASE.

MARRIAGE AND DISEASE.

A STUDY OF
HEREDITY AND THE MORE IMPORTANT
FAMILY DEGENERATIONS.

BY

S. A. K. STRAHAN, M.D.

BARRISTER-AT-LAW,

MEMBER OF THE HONOURABLE SOCIETY OF THE MIDDLE TEMPLE, LONDON
MEMBER OF THE COUNCIL OF THE MEDICO-PSYCHOLOGICAL
ASSOCIATION OF GREAT BRITAIN AND IRELAND;
MEMBER OF THE MEDICO-LEGAL SOCIETY OF NEW YORK,
MEMBRE DE L'UNION INTERNATIONALE DE DROIT PÉNAL, ETC. ETC.

"There is a destiny made for a man by his ancestors, and no one can elude the tyranny of his organisation."—MAUDSLEY.

NEW YORK
D. APPLETON AND COMPANY
1892

Authorized Edition.

CONTENTS.

MARRIAGE AND DISEASE.

INTRODUCTORY.

THE doctrine of the hereditary transmission of family characters, pathological as well as physiological, although now incontestably established, has by no means been accorded the general recognition its great importance so clearly demands. With our present knowledge, there cannot be the very slightest doubt in the mind of any one who has even casually considered the subject, that much of the disease, both physical and mental, which afflicts this and every other civilised people on the face of the earth is to a large extent the result of hereditary transmission of a degenerate constitution or predisposition to disease, brought about by the deteriorating influences of civilised life. Nor can it be doubted that the tendency of the age is toward the cultivation and spread of these hereditary diseases; for while our modern, exciting, feverish, highly artificial, mode of life is prolific of disease and degenerative changes in the organism, the customs of civilised society, as at present constituted, are designed to bar the course of Nature, and prevent, so far as is possible, the operation of those laws which weed out and exterminate the abnormal, diseased, and otherwise unfit in every grade of natural life.

The beneficial working of these natural laws is to be seen among savage and less civilised peoples, where the mode of life is less artificial than our own—that is, where the animal is in more perfect accord with his environment than is the case among the more highly civilised communities. Even here we find the unfit ; for the unfit is a variation, a pathological variation, and variations both pathological and physiological must of necessity at times appear, even under the most favourable conditions. But when such variations from the normal or healthy type do appear in natural life, their survival is of brief duration. Here, there is no continuance of the deformed, the crippled, and the feeble. Natural selection remorselessly weeds out all individuals who from any cause are unfitted to their natural environment, and it is this law which maintains the high standard of health which exists among all savage and semi-civilised races.

How different is this from what obtains in the highly artificial life which civilised man has built up or created for himself ! Here the weakling, the cripple, and the diseased, who in the natural life would at once succumb, are nursed and protected ; they are surrounded with an artificial environment designed to render a continuance of life possible, and, finally, if they be endowed with the procreative function, they are permitted to call into existence a wretched offspring. In this way we make an attempt to hold Nature at bay. We fight and struggle with all our strength against the inexorable law which condemns the unfit to extinction. Fortunately for the race, our success, our greatest success, can only be temporary. At best we can only for a little time put off the evil day, if it can be called evil, and where in the end is our gain? In the more primitive and natural conditions of life, the weakling is at once removed, because of his inherent weakness, his unfitness, his inability to suit himself to his surroundings; whereas we, in our wisdom, endeavour to postpone that consummation, and it is not until one, two, or perhaps three generations of suffering wastrels have fretted and wept their hour upon the stage, that we stand aside, unable longer to bar the path, and see Nature do her work.

I do not for a moment intend to question the righteousness of these endeavours of civilised man on behalf of his afflicted brother. With that great question we have nothing here to do; we can but admire the beauty of the unselfish and Christian spirit which prompts his action, and regret that Nature vouchsafes him such a sorry reward. Our business on this occasion is of a far more practical character than the consideration of such vexed questions. It is to tell those whom it may concern—and it concerns the whole race—that disease is being handed down from father to son, from mother to daughter, from parent to child; to point out how, all-important to the race, how inexorable and unescapable is this law of hereditary transmission of disease, or of liability thereto.

Sir James Paget has said, " If one could set before one's self the gravest and most important problem in all pathology, it would be that which concerns the inheritance of disease ; and, as Sir William Gull has rightly stated, the inheritance not of disease alone, but of that which from generation to generation shall obliterate the disease which one ancestor may have acquired." *   This is undoubtedly true, and until this most important problem is more deeply studied and more clearly understood, it is certain that the physician will not be able to exercise to the full his highest function, which is not to cure, but to prevent disease.

At present the public appear to know little of this law of hereditary transmission as applicable to themselves, or, if they know it, they ignore it.   For while we are most careful not to transgress this law of Nature in the breeding of our horses and cattle, and even our dogs and cats, few of us appear to give a moment's thought as to what may be the physical, moral, or mental inheritance of our children.   This disregard must arise from either ignorance or carelessness, and it is the duty of the physician to make it impossible for any man to plead the former.   Surely in these days of almost free education, when the elements of physiology are taught in every school, it should not be difficult to impress upon the minds of boys and girls the fact that this law of hereditary

* Address on Collective Investigation of Disease.

transmission applies to all Nature's creatures, to the highest as to the lowest, to man as surely as to the inferior animals. Young men and women should be told of those diseased conditions, as insanity, epilepsy, scrofula, drunkenness, which are more certainly transmitted from parent to child, and be impressed with a lively sense of the terrible responsibility resting upon those who, themselves bearing such brand of unfitness, continue their kind, bringing immense suffering upon the earth, which would never have existed had they exercised discretion and self-denial.

When this is done, when men and women can no longer plead ignorance, and are able to appreciate even in part the gravity of these questions, it is surely not too much to hope that some at least will pause before calling into existence creatures foredoomed to sorrow and suffering and ultimate extinction.

I grant that most attempts at interfering with the instincts of man have proved futile. In these matters passion and desire are permitted to dominate reason in the vast majority of cases; but that is no reason or excuse for silence. The truth concerning the tremendous issues at stake should be proclaimed, though we know the great many will ignore our most earnest warnings. The duty of the teacher, which is to instruct and forewarn, is not the less clear because of the unwillingness of the people to hearken to the truth. I have been told that teaching upon this subject must prove absolutely fruitless; that men and women will be led by their instincts in the selection of husbands and wives, and will refuse to be guided in any degree by the teaching of the physiologist; but this, which, even if true, is no argument against teaching the truth, I decline to believe, for the simple reason that I have known good and honourable men and women who, aware of some grave hereditary tendency to disease within themselves, have voluntarily abjured marriage and let their grievous legacy die with them. The case is not, therefore, without hope, and I cannot but believe that there are many thinking men and women who, if they only understood what a terrible heritage their children *must* take with their breath of life, would cheerfully accept the inevitable, and, renouncing all the pleasures and pride of the matrimonial state and paternity, let their miserable estate lapse at their death for want of an heir.

I do not believe in the generally accepted cry that men and women are blindly led by their instincts in these matters. The matches made every day disprove it. Love is not, after all, so overpowering a passion that it cannot be guided by reason, nor is Cupid so blind as he is painted. For rank or wealth a man will woo, a woman give her heart, or at least her hand; and this being so, surely where the reward is so infinitely greater, where the whole future of the coming generation is at stake, rational people will not permit their passions to run riot and overbear their reason.

But besides these extreme cases in which the hereditary taint, the predisposition to disease, is so decided that marriage should not be thought of, there is the still larger class of those in whom the taint is so mitigated, that, with a properly selected partner, a fairly healthy family might be reared, and to this very large class instruction is of the utmost importance. These must be taught that their only safety depends upon their selection of partners, and that it is only by strict attention to this that those bearing within their nature the germ of hereditary disease, or tendency thereto, can hope to be represented in remote posterity. If such a tainted one choose a partner from a family free from his own taint, or, better still, free from any marked hereditary taint, all may be well; but if he marry a person having the same abnormal bent, belonging to the same distorted offshoot from the normal stem to which he himself belongs, then disaster will surely fall upon the luckless children called into being.

It is a question how far the present evil state of things should be allowed to go before the strong arm of the law should interfere. At present, save only the idiot and the raving maniac, who are in the eye of the law unable to make a contract binding on themselves, there is no one so diseased, crippled, or deformed that he or she may not marry, and become the parent of a suffering, helpless family, so far as the law is concerned. Even if, during a "lucid interval," a lunatic contracts a marriage, it is valid, unless at the time there happens to exist an unrevoked commission of lunacy. Can any one assert that this state of things is for the good of the commonwealth?

Many high authorities have expressed the opinion that those suffering under gross hereditary disease, or tendency thereto, should not be permitted to continue their like, and so contaminate the race, but I fear the day for such legislation has not yet come. Moreover, I think it only fair to assume that much of the present continuance of transmitted disease is the result of ignorance on the part of the people, and, on this assumption, some effort should be made to educate them to a knowledge of how terribly relentless and unavoidable is this law of Nature, before calling upon the Legislature to interfere in what might be so much better done by public opinion and individual effort.

Much might also be done by pointing out how some of these tainted constitutions may be acquired *de novo;* how the man or woman whose family has a clean bill of health can by wicked and vicious habits build up insanity, or epilepsy, or phthisis, or gout, etc., to be handed down to posterity, and how other diseases may be acquired which shall have a terrible effect upon children afterwards begotten. It should also be taught how a man or woman with a bad family history may, by a steady and virtuous life, a strict observance of the laws of health, and proper care in the selection of a partner, live down the evil, so to speak, and leave an unencumbered estate to the children of the next generation.

When these things have been taught and found ineffectual, but not till then, should the Legislature be called upon to interfere, except in those cases in which the drunkenness, disease, or crime is so ingrained in the nature of the individual that no amount of care or forethought could be expected to give the children what might be called " a reasonable chance." Among these latter would be included imbeciles, confirmed epileptics and drunkards, those who have been insane more than once, and habitual criminals, all of whom should be at once denied the right of procreation.

CHAPTER I.

HEREDITY.

HEREDITY is that mysterious influence which fore-ordains that the offspring shall be in the likeness of its parents. It is one of Nature's great fundamental laws, and it is universal. In the meanest forms of animal and vegetable life, as in the highest, each family must produce its like. Grapes are not gathered from thorns, nor figs from thistles. All through Nature, from the amœba to man, from the yeast-plant to the oak, every kind produces after his kind. Yet plants and animals are not tied down by this law to endless sameness. Certainly Nature will not permit any rude violation of this law of hereditary transmission, but, where the change is sufficiently gradual, where it is, so to speak, the work of Nature herself, it is with the aid of this very law that modifications are built up, the changes brought about in the individual by the action of the environment being repeated and perpetuated in the offspring by the action of the laws of heredity. These modifications sometimes tend toward the elevation of the family type—evolution, and sometimes toward its extinction—dissolution.

The influence of this natural law begins with life and only ends with death. The life of the individual begins when a single cell from the male (the sperm) meets with a single cell from the female (the ovum), and this combination calls into being a new creature. From this instant heredity is at work. That the new creature thus produced should invariably grow up in the gross and outward likeness of its parents is as strange as it is at present inexplicable. But what is still more wonderful is the fact that these two germ-cells, these two microscopic masses of apparently homogeneous protoplasm, which convey from parents to offspring the *racial* peculiarities of the parents, have the further power of transmitting to the children the various *individual* peculiarities of the parents, as length of limb, colour of hair, cast of features. Nor does the marvellous stop even here, for these potent atoms almost invariably convey to the offspring, as seen in the human family, such infinitely complex and subtle similarities as trick of gait, tone of voice, longevity, liability to certain diseases and immunity against others, together with mental qualities, and even moral bent.

The child takes his life from his parents, and with that life he takes a certain estate made up of moral, mental, and physical characters. This estate *must* be entered upon however encumbered; he is the *heres necessarius* of his parents; he cannot renounce his claim upon this estate and let it pass on to some other heir, neither can he alienate his life-interest therein. He takes it with his life, and only with his death does his

interest in it lapse. Of what vast importance is it,
then, that this estate should come to the child " free
from waste and dilapidation ; " yet how often it proves
a veritable *hereditas damnosa* we have but to look
around to learn.

As might be expected, many attempts have been
made by science to explain this wonderful law which
governs the growth and development of germ-cells, and
enables them to convey not only the gross racial traits,
but the most minute and subtle individual characters
from parent to offspring; yet, although some of the
greatest minds of our age have wrestled with the subject, no one has broken the secret-house. Darwin ela-
borated the hypothesis which he called " Pangenesis "
to explain how the germ-cells gained this extraordinary power. He supposed that minute bodies, which
he called " gemmules," were thrown off by all the cells
of the body and congregated in these germinal cells.
These gemmules, it was supposed, had the power of
reproducing cells similar to those from which they
came, and so the germ-cells were enabled through
these gemmules to produce a body complex and identical in every particular with those from which the
germ-cells came. But, unfortunately for science, this
is but theory; of these potent gemmules we know
nothing; they have not even been proved to exist. All
we have is the ingenious effort of a great mind to
fathom what at present seems to be the unfathomable.

Nor has any other searcher discovered more. Herbert
Spencer's " physiological units " and Häckel's " mole-

cular motion in cells" are equally hypothetical; while nothing is really known as to the action of Weismann's "germ-plasm" or Nägeli's "idioplasm." We have no evidence that the protoplasm of the reproductive cell differs from that of any other cell; yet that it has the power of transmitting from parent to offspring most subtle peculiarity of mind and body we know from experience, but how or why it should be so endowed still remains one of Nature's profoundest secrets.

Notwithstanding the fact, however, that we cannot follow Nature in all her mysterious workings, our course is perfectly clear. We know that "like produces like" all through Nature, and it is our duty to recognise the potent influence exerted by this law in determining the conformation of mind and body in the human family, and endeavour to use this knowledge for the benefit of the race. The hereditary transmission of physical characters has been known from the earliest times of which we have any record, and man has benefited by this knowledge in the breeding of animals from time immemorial. (He has not hesitated to apply it even to the human family when the excellence of the offspring was of pecuniary interest to him, as in the case of slaves.) Yet, for some inscrutable reason, when it becomes a personal question he ignores this great fundamental law, and every year thousands of children are begotten with pedigrees which would condemn puppies to the horsepond.

What, I would ask, would be thought of the man

who proposed to increase his flock or herd from animals
whose ancestors had suffered or died one after another
from some common disease ?   Who is he who does not
choose his finest animals for breeding purposes, and if
an otherwise superior animal have a fault, endeavour
by judicious " crossing " to lessen, and in time eradicate,
that fault ?   But while we thus study with jealous care
what shall be the natural inheritance of our horses,
cattle, and hounds, we give not a passing thought as
to what may be that of our children.   Animal passion,
sickly sentiment, or the desire for rank or wealth, is
permitted to jostle aside our reason, and even some of
us whose business in life it is to improve the breed of
some useful animal by careful selection may be found
at the same time rearing a family of children from a
mother who is a member of a family saturated with
disease.   While we practically ignore this law in relation to the human family, we assuredly never forget its
great influence in the case of the beasts that perish.
Why ? Is not man worth more than many brutes ?

Of this refusal, this obstinate refusal on the part of
man to recognise the effect of this law of hereditary
transmission upon the human family, Dr. Maudsley
observes : " Because it has been the fashion to look
upon an individual as if he were the product of an
independent creative act and a self-sufficient being—
because men commonly look not beyond a single link
in the chain of causation—therefore it has been impossible hitherto to uproot the erroneous notion, explicitly
declared or implicitly held, that each one is endowed

by Nature with a certain fixed mental potentiality of uniform character. But now that observation reveals more and more clearly every day how much the capacity and character, mental and bodily, of the individual is dependent upon his ancestral antecedents, it is impossible to deny that a man may suffer irremediable ill through the misfortune of a bad descent. Each one is a link in the chain of organic beings, a physical consequent of physical antecedents; the idiot is not an accident, nor the irreclaimable criminal an unaccountable causality." *

As I have said before, men and women must be made to understand that this law of hereditary transmission of health and disease, both mental and physical, applies equally to all Nature's creatures, and that if the race is to be improved, it must be done on exactly the same lines as are followed in the world of inferior animals, viz., to cultivate the good, modify and improve the indifferent, and let the absolutely bad die out. "Like father, like son," is an old saying and a true one. Heredity is a law from which there is no escaping. Our bodily and mental development, as received from our ancestors and modified for better or worse by ourselves, is a certain heritage for our children. As we improve our condition mentally or bodily, so will our posterity be gainers, and as we degrade our natures, so shall our children suffer degradation. Like begets like, and whether the particular bent in the parent be for good or evil, toward health or disease of

* "Physiology and Pathology of the Mind," by H. Maudsley, M.D.

mind or body, it will to a certainty leave its mark upon
the offspring.

The fact that physical characters are transmitted no
one denies, and the researches of Mr. Galton and others
prove conclusively that quality of mind is as frequently
and certainly handed down from parent to child as are
physical peculiarities.   When do we read the biography
of a man, and not learn from what ancestor he inherited
this particular mental or moral character, and from what
other that ?   As the scrofulous beget the scrofulous,
and the gouty the gouty, so do the neurotic beget the
neurotic and the insane the insane, the immoral the
immoral and the criminal the criminal.   The child
whose ancestors have been criminals and jail-birds
takes as naturally to crime as does the sheep-dog to his
duties on the hillside.

As man's individuality—that is, his own peculiar
conformation of mind and body—is in all cases the
outcome of many generations of building up, so it
must be in all cases the work of generations to eradicate any well-marked character, or otherwise modify
the family type.   Yet in every case the type can be
modified—for evil only too easily, for good by wise
marriages and scientific medical treatment persisted
in.   Of course cases are to be found on every hand
where reversion to the healthy is impossible, where the
individual is so degenerate, so far removed from the
normal, that, despite outward appearances, the necessarily fatal type has already been reached.   In some
of these cases, as in the impotent and sterile idiot and

cretin, it is evident to the world that Nature has put down her foot and refused to go further; but in many others, although the necessarily fatal type has been attained, it is not discoverable, and it is not until, after one or more marriage, we find them childless, or have seen their wretched children, one after another, succumb long before reaching the procreative period, that we recognise the fact that the end has been reached. " When men wilfully frustrate the noble purposes of their being, and selfishly ignore the laws of hereditary transmission, Nature takes the matter out of their hands, and puts a stop to the propagation of degeneracy."

I will give a case in point. It is the history of the family of a patient of my own, and shows how, when a certain low-water mark of vitality is reached, Nature rids herself of the useless :—

J. W.'s FAMILY.

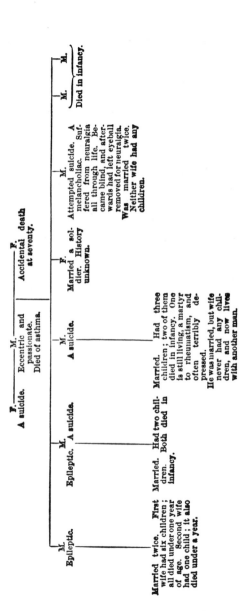

M. Epileptic.

M. A suicide.

F. A suicide.

M. Eccentric and passionate. Died of asthma.

F. Accidental death at seventy.

M. A suicide.

F. Married a soldier. History unknown.

M. Attempted suicide. A melancholiac. Suffered from neuralgia all through life. Became blind, and afterwards had left eyeball removed for neuralgia. Was married twice. Neither wife had any children.

M. **M.** Died in infancy.

M. Epileptic. Married twice. First wife had six children; all died under one year of age. Second wife had one child; it also died under a year.

M. A suicide. Married. Had two children. Both died in infancy.

Married. Had three children; two of them died in infancy. One is still living, a martyr to rheumatism, and often terribly depressed. He was married, but wife never had any children, and now lives with another man.

Now the degenerate father of this wretched family most likely owed his arrival at maturity to artificial means. In natural life such a creature, passionate, eccentric, and the brother of a suicide, would in all probability have succumbed to his innate unfitness. We do not know whether, or how often, he was prevented by others from carrying out "Nature's remedy," which his sister in a fortunate moment applied, but that his unstable temper and eccentricity would have proved fatal at an early age in natural life we are justified in assuming. That he did not achieve the "consummation devoutly to be wished" before maturity, or was not prevented begetting his kind, was an unmitigated misfortune to himself and to the world. By his life, and the lives of his unfortunate children, the world gained nothing. By his early death or enforced celibacy, what suffering would it have escaped! But my reason for bringing forward this family history at present was to demonstrate how infantile mortality and sterility are used by Nature to stamp out the unfit. Of this miserable man's six sons, two were happily carried off in infancy. Each of the other four lived to beyond sixty, and all were married. They had amongst them no less than six wives, and there were born to them twelve children; yet to-day their last and only representative is a wretched, crippled melancholiac, whom Nature has branded with sterility.

Thus do all hereditarily transmitted diseases, or, more correctly, degenerations, tend to one goal—extinction of the family; and it is only by judicious marriages,

begun before the degeneration has become too deep and
persisted in for generations, that a blighted family can
rise to the level of health and hope to live in posterity.
In the family cited above an attempt was made to
continue an unfit variety of the human race, with what
result we have seen, viz., a refusal of Nature to continue
the variety. Of course it is only where the natural
laws are interfered with, as they are in civilised life,
that such could have occurred. In the natural state
this family, in all likelihood, would have been stamped
out two generations earlier, and the nineteen descendants, whose lives were but a cheerless procession to
the grave, would never have existed.

In natural life the necessarily fatal type is reached
much earlier than under civilisation, and consequently
the standard of health is distinctly higher among bar-
barous than among civilised peoples. Even such an
apparently innocent abnormality as, say, colour-blindness cannot be cultivated in the natural state, for the
first individual bearing the degenerate character would
undoubtedly fall a victim to some enemy because of that
character, and so the variety would be lost. Of course,
in civilised life the necessarily fatal type is ultimately
attained in every case where reversion does not take
place, in spite of all man can do to stay or prevent that
consummation. Thus Nature ultimately rights herself
in all cases by setting her veto upon the perpetuation
of disease, and were it not for the suffering experienced
before oblivion is reached, nothing need be said; the
erring might be left to their fate. But this suffering,

which is serious even in natural life, becomes grievous in these days of the higher civilisation, when those who owe their continuance in life to the exertions of others are not only permitted, but are aided by every device known to science, to propagate their kind. Now-a-days, when the maniac, the melancholiac, and the would-be suicide of yesterday, the imbecile, the epileptic, and the habitual drunkard are married and given in marriage, the suffering has become so terrible, the contamination of the race so great, and the care of the useless offspring begotten so heavy a charge upon the community, that if some effort be not made voluntarily to stay this curse upon the land, the Legislature must be called upon to interfere.

As I have already said, excepting only the idiot and the raving maniac, who in the eye of the law are unable to make a contract binding on themselves, there is no one so diseased, crippled, or deformed that he or she may not marry and become the parent of a suffering, helpless family, so far as the law is concerned. That this should be so is a scandal upon our boasted civilisation. Why should the industrious citizen, who for years worked hard and saved money that he might marry with some reasonable prospect of being able to support his family when it came, be called upon to support the helpless, worthless offspring of the drunkard, the imbecile, the criminal, and every other wastrel who chooses to become a parent ? But this great question hardly comes within the purview of this work, and, notwithstanding its fascination, we must leave it for the present. Speaking, however, from a purely scientific standpoint, I will assert, without fear of contradiction from those able to judge, that legislation tending to limit the propagation of the insane, the epileptic, the drunken, the criminal, and the pauper would have a markedly beneficial effect upon the health and comfort of our people. As Dr. Benjamin Ward Richardson has wisely said, "The first step towards the reduction of disease is, beginning at the beginning, to provide for the health of the unborn. The error commonly entertained, that marriageable men and women have nothing to consider except wealth, station, or social relationships, demands correction. The offspring of marriage, the most precious of all fortunes, deserves, surely, as much forethought as is bestowed on the offspring of the lower animals. If the intermarriage of disease were considered in the same light as the intermarriage of poverty, the hereditary transmission of disease, the basis of so much misery in the world, would be at an end in three, or at most four generations." *

Yet, notwithstanding all that has been said of the relentlessness of this law of heredity, the unfortunate inheritor of a poorly or viciously developed mind or body is not to fold his hands and say, "It is useless for me to strive." Not a single passage in this book is intended to support such a view. Let such unfortunate work, and work patiently in a good cause. Although he might as well hope to rid himself of his shadow as of any deeply marked hereditary tendency,

* "Diseases of Modern Life."

whatever its most prominent traits may be, he may by strenuous effort, by judicious treatment, and a pure and healthy life, do much to bring himself nearer to that grand ideal of manhood which none of us approach too closely. Nor can any man be engaged in a nobler or holier work than purging the body of the sins of its fathers.

This, then, is heredity. This mysterious, unknown, and possibly unknowable something which moulds the child after the fashion of its parents. It is this law which has made man what he now is; it is this law which shall make man what he may yet be.

CHAPTER II.

VARIATIONS.

So far we have been speaking of heredity broadly, as it affects the race generally, and as if there were no other influence at work in moulding the offspring than this "like father like son" rule. Certainly there is no other influence which exercises a tithe of the power which heredity pure and simple does in foreordaining what the child shall be; but if we consider for a moment, we shall see that there must be other influences at work. If it were not so, if heredity were not in any way interfered with, the child must, of necessity, be a perfect mean of the parents, and all children of the same parents must be identical. Now we know that this is not so. An exact likeness, physical, mental, or moral, is never transmitted by inheritance; such a thing is impossible. It has been said that no two blades of grass are exactly alike, and it is certain that no two faces, bodies, minds, or moral natures are exactly alike. Each person is endowed with a certain individuality which distinguishes him from all others. At times we do meet with a case in which the child is, to a marvellous degree, a reproduction of one or other parent, in feature, form, and limb, or perhaps in mind, or even in both; but such cases are extremely rare, and in no case must we ever hope to find a *perfect* likeness. Nature does not slavishly follow any one type or pattern, but revels in infinite variety within certain limits.

The slight variations constantly met with in the family are due, for the most part, to the various blendings of the parental characters, which a moment's consideration will show may be endless. Remnants of the countless characters of the ancestors are present in each parent, some strong, some weak, some standing out prominently, others almost effaced. Nor are they even thus a constant quantity, for while the life of the individual develops one, it may allow another to fade almost to oblivion. Thus the children begotten at different periods of life, even if they were mere examples of the mean of the parents, must vary considerably. As it is, one child will inherit some peculiar character from one parent, in whom that particular character is just then prominent and active; another child will inherit largely some other character from the same or the other parent, while a third child may, by some happy blending of perhaps mediocre parental characters, become the fortunate inheritor of some physical or mental character of a high order; or, conversely, by some unlucky mischance, parental characters, good in themselves, may combine to form a compound markedly bad.

The inheritance of the child is a piece of patchwork, a thing of shreds and patches. It is a mosaic made up of innumerable characters, all varying in form and tint, and changing with every generation as some drop out and new ones take their place. The variously coloured pieces are so numerous that even Nature cannot arrange them twice in exactly the same order. But although they are differently arranged in every case, each family has what might be called a "family pattern." In every member of the family the tracery and tinting may differ in parts, yet there is a common resemblance which is seldom wanting. This "family pattern" is being constantly modified by the introduction of fresh characters, which in the course of time may change the pattern altogether; but such change is always very gradual. In the ordinary course there is no rude interference with the general arrangement, which maps out the great fundamental curves and grouping which give a common likeness, and make children of the same family like, however unlike. Occasionally such radical changes do occur, and the graceful curves of health may be distorted to the angles of disease. The beautiful tintings which represent mental and moral worth may be replaced by glaring, ill-placed patches of colour, or all order may be lost in the chaos of idiocy; or, again, the pattern may take on the beauty of genius. But no such arrangements continue long. They never become fixed family characters. In one or two generations the old family pattern—modified, it may be, but still the old pattern—will appear again, or the pavement will take on the blankness of death.

This blending of the parental characters, which we see is capable of endless variety, is often very considerably fettered by the fact that all characters are not equally potent; generally speaking, those more recently acquired are not transmitted with the same certainty that those of long standing are. These latter are said to have gained a "prepotency" by long descent; by repeated transmission they have become fixed and prominent characters in the family, and the presence of one such character in a parent tends to materially limit variety in the offspring. But we shall consider this subject later on, when we have learnt something of how "acquired characters" affect the family.

Again, it should be pointed out that what are commonly looked upon as most striking variations in the family, are often, in reality, not variations at all, but reversions. In such cases the branch of the family to which the parent belongs has, in consequence of some combination of circumstances, deviated from the family type. With each generation the divergence has increased, but it has not gained any fixity, and with the infusion of fresh blood the offspring "throws back" to the original family type. It is looked upon as a variation, whereas it is a fair representative of the family, and it is the parent who is at fault. But this subject of reversion we shall also postpone for a short time, while we endeavour to see what effect is exerted upon the race by the environment—that is, the action of the whole outer world.

It would perhaps be convenient to point out here that all variations from the normal standard of development and health can be classed under two heads, physiological and pathological:

Physiological where the variation takes the form of special development in any healthy direction, whether physical, mental, or moral; and

Pathological where the variation tends downwards to degeneration and disease, or to absence of due development, as where insanity, idiocy, bodily deformity, gout, scrofula, and like hereditary diseases and imperfections brand the family stock—that is, where the variation tends toward dissolution.

And now let us briefly consider the action of the environment in the production of variations. The influence of the environment is at work from the instant of conception. The child's environment, made up of pressure, food, air, exercise, education, association—the whole outer world, in fact, is with all Nature ever changing; and as it is impossible that this ever can be identical in any two cases, so it is impossible, on this ground alone, that any two children ever can be exactly alike. Thus each individual must be a variation from the mean of his parents, as he must also, and from the same cause, be a variation from the normal. (The normal is of necessity ideal, as we cannot point to any individual and say, "This is the normal standard," for the reason that no one is, perhaps, absolutely healthy, and every one has been, and is being, modified by his environment.) But although all persons are thus variations from the normal stock because of the action of the environment, it is not necessary to look upon them as such. For all practical purposes the average of the healthy mass may be taken as the normal standard. Practically these finer variations are of no moment, and having recognised their existence and cause, we may pass on to the consideration of the more gross. It is only in those cases in which the individual has strayed so far from the normal as to be "out of the common" that we are asked to look upon him as a variation, and consider whether he is to be classed as physiological or pathological—whether he is on the way to a higher development, or on the down path to extinction.

This action of the environment upon the individual has not been given the attention it deserves in this connection. Some writers refuse to recognise it as a cause of variation, believing that its action affects the individual only slightly, and does not modify the family type through him; but we shall find, upon inquiry, that it is one of the most potent influences at work in the changing of family and racial types.

Man, as we know him, is a creature of circumstances; he is moulded by his surroundings. His condition of mind and body, in whatever position we find him, is the result of the action of the environment upon him, and as his surroundings are ever changing, so is he ever changing with them. No man is mentally, morally, or physically exactly what he was even a year ago, and no one will aver that this change is not

in great part the result of contact with the outer
world. No man can cut himself off from the moulding influence of surrounding Nature, of which he is a
part, and the impressions received from it will affect
not only him but his posterity. When Jacob modified
the environment of his flocks by fencing them around
with white rods, he did not do so for the purpose of
affecting the flocks alone, but that he might profit by
its modifying influence upon the generations to follow.

But this is a subject for demonstration. Let us take
a case and see how this influence works. Suppose
we take twin brothers, who, notwithstanding the in-
equality of environment they have experienced in
pre-natal life (as difference in position and pressure,
size and arrangement of placentæ, &c.), have entered
the world as like to each other as may be. Now,
send one of these infants to a healthy farmhouse to be
brought up a ploughboy, and let the other be reared
in the back slums of a great city in the midst of
poverty and vice, and what will be the result? It
will be this : that the one who breathes the pure air
of heaven, feeds on a plain but healthy fare, and does
an honest day's work with his muscles every day, will
at least develop the physical part of his nature, and
reach manhood full of health and strength, and experience a delight in living ; while his brother, bred
in the lanes of a great city, seldom, if ever, breathing
other than a vitiated atmosphere, and subsisting on
food wanting in many of the essential constituents of
a healthy diet, will arrive at manhood—should he

reach that stage—wan and pinched, no more like his twin brother in the country than Hamlet was to Hercules. This is the action of the environment upon the individual. Here we see its effect upon the physical organisation during the early years of life, and no proof is needed that its effect upon the mental and moral natures is equally, if not still more powerful.

But let us follow these brothers. Suppose the ploughman, full of life and vigour, marries a strong, healthy country-woman, while his town-bred brother marries her sister, who, like himself, has been brought up in poverty in a large town, and whose store of health is more or less reduced in consequence; can we for a moment suppose that the natural inheritance of the children of these twin brothers and these sisters will be identical? The one family, having inherited the sound physical constitution of the ploughman and his healthy wife, will grow up vastly different from their stunted, pale-faced, town-bred cousins, who show the second step on the downward path leading to the extinction of the family: different as is the typical yeoman from the typical Cockney, who, it has been said, " has no grandfather." *

This is the action of the environment upon the

* Mr. Cantlie, after prolonged and careful search, could not find a single person whose ancestors, from the grandparents downwards, had been born and bred in London. He describes some miserable creatures who most nearly approached this record, and then remarks :—" I have never come across the children of any such, and I believe it is not likely I ever shall. Nature steps in and denies the continuance of such."—"*Degeneration amongst Londoners.*"

physical organisation, and as it acts there, so does it in the mental and moral worlds. As the muscular system and health can only be developed and preserved by good food, air, and exercise, so the mental faculties can only be enlarged and brightened by education and example; and as surely as the blush of health fades before starvation and disease, so does moral loveliness fade in the presence of vice and degradation. "Train up a child in the way he should go, and when he is old he will not depart from it," is as true of the physical and mental as it is of the moral nature.

These, then, are the causes of variations in the family, viz. :—

1. The various and unequal blending of the parental characters, which possibly, and in some cases certainly, depends upon the humour, or mental or bodily condition, of the parents when they become such.

2. Reversion, or "throwing back" to a family type lost some generations before; and

3. The action of the environment upon the child from the moment of its conception.

CHAPTER III.

ACQUIRED CHARACTERS.

"Man is subject to all the general laws of animal nature. The law of heredity is one of those from which he cannot escape, and it is this law, which, under the influence of the conditions of life, fashions races and makes them what they are."—DE QUATRE-FAGES.

FROM the little we have heard of the action of the environment upon the individual, it will be understood that it is not alone the blend of family characters, received from our parents, which we in our turn hand down to our children, but those characters modified, as they may be for better or worse, for good or evil, in our own lives, together with those acquired by habit, occupation, mode of life—by the action of the environment, in fact. We should never forget that we live not for ourselves alone, but for posterity: that we hold the well-being of the race in trust for our children, and should, as honest trustees, let the estate pass on to the next heirs, free from waste or dilapidation when our life-interest in it is done. The man (or woman) who undermines his physical health, or degrades his mental or moral nature, is dishonest. He is robbing the children yet unborn. The spendthrift, who encumbers the broad acres of the family estate, and leaves his children penniless, no more surely robs his children of their rightful heritage, than does the man (or woman) who, by a wicked and vicious life, degrades his nature, thereby making his children physical, mental, or moral beggars.

Some students of heredity deny that characters acquired by the individual are transmitted to the offspring, and foremost amongst these is the German savant Weismann, who has done so much to advance our knowledge on all things relating to heredity. With these, however, we cannot agree. We hold, with the great majority of authorities on the subject, that all characters acquired by the individual have a distinct effect, more or less powerful, upon the offspring of the individual. To admit that these acquired characters are not transmitted, nor transmissible, is to make a clean sweep of evolution. If acquired characters have no influence upon the offspring, from whence came the innumerable characters which are to-day transmitted? Are they each the result of a distinct creative act, and, if not, how did they arise, how did they come into existence? Have they, and every other of the countless myriads of characters which are, and which through all past ages have been, exhibited in all the various human races, been originally and ever present in the germ plasm? Some physiologists assert that this is so, that the germ cell acts but as a torch whose touch lights up the spark of life, and which is passed on from generation to generation, unchanged and unchangeable; but this theory is hardly less difficult of belief than the old theory of the germ cell containing the animal to be produced in miniature. These physiologists hold that the cells possessing these wonderful powers of production, having started the condition we know as life in the new being, separate themselves from the organising mass of the fœtus at an early period, take no part in its development, but simply lie apart, dormant and unimpressionable within it, until the new creature reaches the procreative period, when they wake up and are ready to be passed on to the next generation, there to light up life, and having lost nothing in the operation and gained nothing, again lie dormant until the maturity of that generation calls them forth once more.

This is an ingenious and a pretty theory, but it is, as I have said, difficult of belief. In the first place, these germ cells must, like all other living protoplasm, be nourished. This nourishment is received from the organism in which they lie. We know that everywhere in nature the creature is influenced by its environment. Why should these cells be the only exception? Again, it is clear that these cells which bear such extraordinary potentiality must proliferate to a marvellous extent, for while the amount of protoplasm set aside for this purpose on the formation of a new creature is a microscopic quantity, on that creature attaining the procreative stage equally potent germ cells are thrown off with lavish prodigality, and this is continued through the whole period of procreative life. In the human female the ovaries have been estimated to contain about 72,000 ova, while in the male the vital units produced are simply innumerable. Therefore this magic germ must transmit its powers to thousands of other cells. These other cells are portions of the general organism of the parent, and that these should bear some impress indicative of their origin and former life is not too strained an inference. These masses of protoplasm are part and parcel of the adult animal organism, we know that they are affected as other cells by certain diseases contracted by the organism, and it is not too much to suppose that, bathed in the same fluids and fed from the same blood-stream, they, like every other cell in the economy, should be liable to the action of the environment; and that this changed condition, whatever it may be, should re-appear in the new creatures or organisms of which they are the first cause and foundation, it is but reasonable to expect. As Sir William Turner said in his address before the British Association at Newcastle in 1889, " Those who uphold the view that characters acquired by the soma [=individual] cannot be transmitted from parent to offspring undoubtedly draw so large a cheque on the bank of hypothesis, that one finds it difficult, if not impossible, to honour it."

In support of this theory of the non-transmissibility of acquired characters, it is generally pointed out that there is little or no proof of mutilations in the parent, such as the loss of an eye or a limb, being reproduced in the offspring. This is to a certain extent true. We do not find that the man who has lost his right hand begets children with imperfect hands, nor do we expect to find it. If the injury or mutilation be to some exquisitely sensitive organ, as the brain, we may have it repeated in the offspring, as we do at times find epilepsy acquired by the parent reproduced in the children ; but in the great majority of such injuries the step is too great a one for Nature, who does nothing by leaps and bounds. From the perfect hand, the work of ten thousand generations, to the imperfect hand bearing two or three fingers is too radical a change ; but let the change be sufficiently gradual, and assuredly we shall have the transmitted type changed. See what changes can be brought about in this same member, the hand, within a few generations. Look at the narrow, elegant, small-boned hand of which the aristocratic family, whose members have not been engaged in manual labour for generations, is so proud : compare it with the broad-palmed, large-boned, knotty-fingered hand of the navvy, the " horny-handed son of toil," and tell us whether these hands were born alike. And just as the hand can be modified, so can any other limb or organ, so can the mind, so can the moral nature.

I might give many examples of the transmission of acquired physical characters, but I shall only mention two which are in themselves most interesting. The first is the process of degeneration which is going on among civilised peoples in the little toe. Herr

Pfitzner has made a great number of observations as to the condition of this member, and has found that in 31 per cent. of males and 41.5 per cent. of females the two terminal bones are fused into one, which is very little, if at all, larger than the terminal bone should be. Nor was this phenomenon confined to adults; he found it equally common in children, and even in infants. Whether the process will continue until the fifth toe disappears we cannot say, but, knowing what we do of evolution, there is every reason to think so. The second case which I shall mention, is the gradual disappearance of the wisdom-tooth in civilised man. It is clear that civilised man does not require the strength of jaw and amount of grinding surface he did before he discovered how much jaw-labour could be done by knives and forks and cookery, and as evolutionists we would expect to find in him jaws and teeth less powerfully developed than in his still savage brother. And that is exactly what we do find. In all savage races we find the whole masticatory apparatus, bones, teeth, and muscles, much better, that is, more strongly developed than in the civilised races. More than this, we find that as the jaws develop less fully the last tooth, the wisdom-tooth, is dying out. Mantegazza in a long series of observations found the wisdom-teeth absent in 19 per cent. of members of the lower races, while they are absent in 42 per cent. of civilised mankind. Now both these conditions are acquired, the absence of the teeth and the atrophy of the toe, and how they are to be accounted for otherwise than by hereditary transmission I am at a loss to understand.

Under changed conditions a plant, an animal, or a man will change. There must be harmony. Nature will not long tolerate a discord. The creature must maintain the equilibrium between himself and his surroundings somehow, or cease to exist, and he can only do this in one of two ways, either he must suit himself to his surroundings, or so modify his surroundings that they may suit him. This latter man frequently attempts, but he can at best only partially carry it out, and so to regain the equilibrium he must himself change. Examples of this abound. It is seen when a people are transferred from one climate or part of the globe to another. In such cases changes are very soon brought about in the people, and they develop characters which were not present before their translation. At first these recently acquired characters are not at all firmly fixed, and there is great liability in the offspring to reversion toward the original family type, but with each generation the new characters become more deeply marked, more intimately ingrained in the organism, and the liability to reversion being gradually lessened, the new characters become fixed and constant in the family. We have a good example of this in the Yankee, for, as Sir William Turner says, "Most of us can distinguish the nationality of a citizen of the United States by his personal appearance, without being under the necessity of waiting to hear his speech and intonation."

Again, no one will deny that the appetite of the drunkard is an acquired character, and cannot he transmit his accursed appetite to his children? Cannot he drink himself into epilepsy or insanity, and afterwards beget children who shall inherit his shattered nervous system, just as the young of the guinea-pigs in which Brown-Séquard had artificially induced epilepsy inherited that diseased condition? Cannot the compositor or the dressmaker, by over-work in ill-ventilated rooms, by want of pure air and healthy food, develop a predisposition to phthisis; and shall we expect the children of such to escape scot-free and inherit nothing of their parents' acquired degenerate condition?

If acquired characters cannot be transmitted, as some say they cannot, how are we to explain the degeneracy and early extinction of the poor families of our great cities, where, in three or four generations, poverty, starvation, and dirt modify the family to extinction? On this subject Dr. Maudsley says :—
" Over-population leads to deterioration of the health of the community by overcrowding and the insanitary condition of dwelling-houses which it occasions in towns. Not fevers only, but scrofula, perhaps phthisis, and certainly general deterioration of nutrition are thus generated and transmitted as evil heritages to future generations; the acquired ill of the parent becomes the inborn infirmity of the offspring." As we have already heard, Mr. Cantlie failed after prolonged and careful search to find a single person of the poorer

classes whose ancestors, from the grandparents down, had been born and bred in London. How was this? Can it be explained otherwise than by the gradual extinction of the family by transmitted physical inferiority or impaired vitality? The healthy labourer going into London loses a part of his vitality because of the wretched conditions under which he exists, and in a few years he is a much inferior man to what he once was. His children inherit his impaired constitution, and in their turn deteriorate a stage further, beget a still more wretched offspring, and so on to extinction. Mr. Cantlie, after describing some miserable specimens of humanity he discovered who nearly approached what he was in search of, says:—" I have never come across the children of any such, and I believe it is not likely I ever shall. Nature steps in and denies the continuance of such."

Thus does the environment create, modify, and extinguish physical characters, and its action is not less effectual in the mental and moral worlds. Just as the degenerate physical development of the poor dwellers in large cities is handed down from parent to child, so has the liability to scrofula, gout, rheumatism, epilepsy, insanity, and drunkenness been acquired and transmitted, as is also the predisposition to crime handed down an heirloom in that degenerate race the instinctive criminal. Just as the fine physical development of the yeoman and the degenerate frame of the Cockney have been the outcome of certain conditions of environment, so have the clear financial insight of the Hebrew,

4

and the mental instability of the neurotic family been developed by ages of habit. " Habit becomes second nature" is an old saying which sums up what we have been preaching, viz., that habit—or, as Shakespeare has it, use—long continued is what builds up our nature. Hamlet says :—

> " Refrain to-night,
> And that shall lend a kind of easiness
> To the next abstinence : the next more easy :
> For use almost can change the stamp of nature."

Darwin says :—" Characters of all kinds, whether old or new, tend to be inherited," and of the truth of the statement there is proof everywhere around. The transmission of acquired characters is well exemplified in some of the inferior animals, as for example, the dog. No one will assert that the sheep-dog, the retriever, or the pointer is the result of an independent creative act. We know that they are sprung from a common stock, although they are now so widely separated physically and mentally. Each one by a certain mode of life has become modified from the original type. The well-bred pointer, that is, one whose ancestors have been trained to the same particular duties for many generations back—will 'point,' as we say, by instinct, the young retriever retrieve to hand after a single lesson, and the sheep-dog take to tending the flock almost of his own accord. Those characters have undoubtedly been acquired by the ancestors of the puppies and been handed down from generation to generation until they have become a part and parcel of their nature. The habits, education, and mode of life—the environment in fact, has brought about changes in the animal's organism, and these changes, which must of necessity be very slight in the first case, are transmitted to the progeny. With each generation the changes are deepened by persistence in the peculiar mode of life, and in the course of a few generations we have a stock of animals which take to a particular work "instinctively." The offspring have not inherited any part of the education of their ancestors, but they have inherited their organisation as modified by their peculiar mode of life: in other words, they have inherited a strong predisposition toward the ways of their progenitors, just as a child inherits a predisposition to the ways of its ancestors. As well might we expect the daughter of the costermonger to take on the modesty and gentleness and tenderness of nature which stamps the daughter of the family noted for these virtues for generations, as to find the qualities of the sheep-dog in the terrier, or those of the greyhound in the bull-dog.

And if this be so—if it be true, as we believe it is —that all characters acquired have an effect upon the offspring, how careful should each one be not to do anything which may leave a stain upon posterity. "The evil that a man does lives after him"—lives not merely as the poet meant it, in the minds of other men, or upon the blotted page of a wasted life. There it does live; but if we wish to know it all, let us read it in the lives of his unfortunate children.

CHAPTER IV.

THIS tendency to revert or "throw back" to the original stock is a wise provision of Nature. It acts as a corrective. It is the genealogical gardener who keeps a watchful eye upon all distorted offshoots, brings them into proper line or lops them off, and so keeps the family tree shapely and free from glaring deformity. Were it not for this law of reversion all those extreme variations from the true family type, however hideous or useless, which from time to time arise from some extraordinary action of the environment or unhappy blending of parental characters, or both, might be repeated and exaggerated indefinitely, even till the true type of the family would be lost in a crowd of mongrels and monsters. Of course the action of the environment would not tolerate this, and so the family would become extinct.

When the individual, from whatever cause, varies so far from the present existing type of his family as to be out of harmony with it, he must also be out of harmony with his environment, for the family type is the resultant of the survival of the fittest, which means those in most perfect harmony with the environment. And as the creature cannot survive long in an environment to which he is not suited, therefore the offspring of the extreme variation from the normal must of necessity return to the original family type, or cease to exist because of its unfitness.

When these extreme variations appear, as they do in a wholly unaccountable manner from time to time, this law of reversion comes into action and is not infrequently successful in bringing back the wanderer; but should it fail in doing so the extreme variety cannot be continued, either there will be no offspring, or if there be it will prove sterile. Thus the continuance of the grossly unfit becomes impossible. When variation is extreme, reversion or extinction must take place, and whether reversion or extinction is to be the verdict of Nature, depends to a great extent upon the conduct of the individual representing the variety. If he or she join with another of the same variety for the continuance of the race, there will probably be no offspring, or if there be it will prove sterile = extinction; while if he or she join with a normal fellow, or one belonging to a different variety, Nature will seize upon the opportunity offered by the infusion of fresh blood, and in the offspring "throw back" to the old family type = reversion. And this law is equally active whether the variation be physiological or pathological. It is impossible to continue a family all idiots, or dwarfs, as it is impossible to continue one all giants, or geniuses. The individual

presenting extreme variation from the normal cannot
continue his like, and were it not for this law of reversion
the offspring of all such must succumb.

Although this law is continually in action, guarding
jealously the family type against gross contamination,
its action is most commonly only observed in those
cases in which the variation from the normal is
marked. Should the variation be only slight this
principle of reversion is less active, the new character
is often transmitted in such cases, and in the course
of some generations may become a fixed and constant
character in the family. In this way new varieties
and races are built up. But should the variation be
extreme, whether recent and unstable, the result of
some inharmonious action of the various influences
at work, or the outcome of several generations of injudicious breeding, Nature revolts against the innova-
tion and throws back to the original.

Of the working of this law of reversion we have
proof on every hand. If we take a person belonging
to some decidedly abnormal variety, as, say, the giant
(physiological), or the deeply phthisical (pathological),
we shall see how it acts. If either of these marry one
belonging to the same abnormal type as himself, there
will either be no offspring, or, if there be, it will show
a marked tendency to perpetuation of the abnormality
of the parents, and this deepening with each generation, the necessarily fatal type will soon be reached
and the family be at an end. Here with the junction
of like abnormal persons, the law of reversion has little

chance of coming into action, and the only other mode of putting an end to an unfit family—extinction —is called upon. Thus the intermarriage of dwarfs or giants is seldom fruitful, while the children of parents both of whom suffer from scrofula, insanity, epilepsy, &c., of a well-marked type are generally carried off in infancy, die before the procreative period is reached, or are sterile. But the result is very different if the diseased or otherwise abnormal one marry a person belonging to some other variety than his own, as the phthisical the robust, or the giant or dwarf one of ordinary stature. Here the law of reversion comes into play, and in the offspring of such unions there is a great effort made to "throw back" to the original stock, that is, in the examples we have taken, the fairly healthy and the mediocre in stature respectively.

Hence we have the rule, that the offspring of individuals of different varieties tends strongly to throw back to the normal, while the offspring of parents belonging to the same variety tends to retain the peculiar characters of that variety, and when the characters become extreme the stock dies out. This rule is well known to breeders of animals, who will tell you that breeding " in and in," that is, breeding from animals belonging to the same variety, will perpetuate, and in time accentuate the peculiar qualities of the variety, be they good or bad, while " crossing," which means breeding from animals belonging to different varieties, will reduce the characters peculiar to either parent, and the offspring will revert to the original stock.

It must here be pointed out that in thus referring to reversion to the healthy type as following the inter-marriage of individuals belonging to different varieties, all degenerate conditions must be taken as belonging to the same variety. The more closely such degenerate conditions as epilepsy, insanity, scrofula, drunken-ness, cancer, and crime are inquired into, the clearer it becomes that they are not only related, but that they are largely interchangeable. For this reason the intermarriage of so apparently unlike temperaments as the cancerous and insane, the cancerous and scrofulous, or the insane and rheumatic, seldom or never result in reversion to the healthy type. Unions of this kind are almost, if not quite, as dangerous to the offspring as those of individuals belonging to families in which the family degeneration has taken exactly the same outward form.

We shall see later that such apparently distinct degenerate characters as epilepsy, insanity, cancer, rheumatism, gout, and scrofula, are in reality but the varying outward signs of a common constitutional depravity, and that they constantly replace one another in succeeding generations of the deteriorating family, and even in different members of the same generation.

The following family history of a patient of my own is a good example of how the outward signs of family unfitness may vary in the different members of the family :—

J. E's FAMILY.

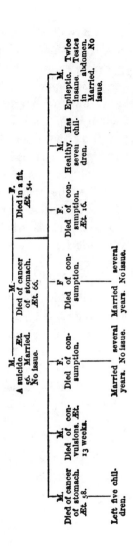

Here, in the family of this cancerous man—whose brother was a suicide—and neurotic woman, we have the innate degeneracy showing itself in the seven children as cancer, consumption, epilepsy, infantile convulsions, insanity, want of development, and sterility. Two of the children happily succumbed before maturity was reached, and of the remaining five three were sterile, two consumptive, one epileptic and insane, and one cancerous. Only one has until now escaped the family blight, but as he may yet develop insanity, cancer, or some other disease degeneration, it cannot be said that even one of the whole family reverted to the healthy type. So far as we know, the taint in the mother was not deep, yet apparently its presence was sufficient to ensure the transmission of the degenerate type by rendering reversion to the normal impossible. Had this woman belonged to a healthy family, there can be little doubt that, notwithstanding the deeply degenerate father, reversion to the healthy would have occurred in some of the children—at all events, the chances of such reversion taking place would have been vastly increased.

Many other cases might be given like the above. That quoted from Dr. B. W. Richardson at page 185 is instructive. There the intermarriage of the cancerous and consumptive temperaments resulted in the destruction of every child by one or other of these diseases.

From what has been said it should be clear that, notwithstanding this law of reversion, it is possible to alter the family type considerably, if only sufficient

time be given to let the new character gain some
fixity. Thus, although we cannot breed a race of
actual giants or geniuses, we can in time materially
increase or diminish the mental or physical stature,
and so long as the modification be not so radical as
to be out of harmony with the environment, there is
little increased danger of extinction; while, if the process be sufficiently gradual, there will be little danger,
after a few generations, of reversion. The truth of
this as applied to physical characters is well known.
Breeders of animals act upon it constantly, and await
the result with perfect confidence. It is not, however,
so generally recognised as applicable to the moral and
mental characters, and for this reason I shall take an
example from the latter class. Let us take the mathematical, or, say, the financial quality of mind, which,
when strongly developed, constitutes a physiological
variation. This variation is often the result of some
happy blending of parental characters, mediocre in
themselves. In such cases it is a newly acquired
character having no fixity. The father or the mother
of the great financier has seldom or never displayed
the mental character which so strongly marks their
child, just as the poet or the orator is seldom the
child of a poet or an orator. If, then, the great
financier marries a woman, brilliant, it may be, in
other ways, but having none of his peculiar quality of
mind, the offspring will in all likelihood revert to the
normal stock, and display little, perhaps very little, of
the father's peculiar ability. Here the new character

has gained no fixity, and, the mother not belonging to the same variety, the offspring "throws back" to the original. Even had the mother possessed something of this mental quality, the result would probably have been the same (though the chances of reversion would have been lessened), for the reason that the stride from the average mental development to that of the mathematical genius is too great for Nature. Had the father's peculiar mental character been less marked, and more especially had it been the outcome of many generations of building up, as it is in the case of the Jew, there would have been very little tendency to reversion, and the children would almost certainly have inherited it. In the Jew the character has become fixed; by repeated transmission through generations it has become a stable and constant family character, and even if the Jew were to marry a woman belonging to a different variety, yet would the children inherit the character. This mental character has been so ingrained in the Jew by continual cultivation, and by intermarriage with those of his own race, nearly all of whom cultivate the character, that it would take many "crossings" to efface it; whereas the same character, even in a much higher degree of development, as it occurs sometimes in the great financiers of other races, is but the outcome of some happy combination of parental characters, and having no stability, we look in vain for its continuance in the offspring.

A point of still greater importance to us, is the fact that this principle of reversion to the normal is as active and efficacious in the case of disease as in other variations—as effectual in calling back the pathological wanderer to the true path as it is in the case of the physiological; and here lies our greatest hope. It is the most cheering fact met with in the study of heredity. Here we have Nature standing by our side, ever ready to assist us in any effort we may make to purify the race. Disease is but "the sins of the father visited upon the children," and if we will but make an effort to clear ourselves of the stain, Nature, ever kind, will go hand in hand with us, and aid us at every step in our good work.

Disease is foreign to Nature—it is an acquired character; and from the fact that when transmitted it so soon, in most cases, attains the necessarily fatal form, it can never have much fixity. Consequently, in all cases in which the disease has not already gained too great a hold, on the infusion of untainted blood reversion to the normal, *i.e.*, healthy, is the rule. In this connection, because of its healing or cleansing effect, reversion has been called the *vis medicatrix naturæ*, and it is of incalculable assistance to us in any efforts of ours to exterminate hereditary disease. But it cannot overbear the laws of hereditary transmission. If the scrofulous *will* marry the scrofulous, and the insane the insane, this benign law is powerless, and the diseased family, going from bad to worse, must become extinct. But if man will not ignore and try to over-ride the laws of Nature, which he can only ignore at his own cost, and which he can never over-ride—if he

will give a tithe of the attention to the laws of hereditary transmission in the production of his children that
he does in the production of every other animal over
which he has control, this *vis medicatrix naturæ* will
assist him, and he may hope soon to have the human
family as free from hereditary disease and imperfection as are the animals in Nature. If man would only
do his part in this great work, Nature could be safely
reckoned upon to do hers.

Let us have a simple example of the working of this
vis medicatrix naturæ. If, in place of the financier,
who was a physiological variation from the normal, we
take the epileptic, or the man of insane temperament,
who represents a pathological variation, we shall find
that the same rule applies. If this man marry a person
of the same abnormal type as himself, that is, a member of some highly neurotic family, the offspring will
inherit the predisposition or liability to insanity more
strongly than that present in either father or mother,
and its chance of passing through life without becoming insane will be markedly less than that of either
parent. But if he marry a woman far removed from
the neurotic or insane type, the children will in all
probability inherit their father's abnormal quality but
slightly, and their liability to madness will be decidedly
less than his own; that is, the children will "throw
back" to the healthy type. And as it is with mental
or nervous disease, so it is with physical. We know,
for example, how the phthisical, and gouty, and scrofulous, and rheumatic temperaments can be and are

increased or diminished in point of gravity at every generation by marriage into healthy or tainted families, and how a disease which has "run in a family" for several generations is sometimes stamped out by a few judicious, if chance, "crossings." Amongst the domestic animals such crossings are judicious, they are the outcome of thought and careful attention, and we know how rapidly effective they are for good. Unfortunately, in the case of man, when they do occur, they are in nearly every case the result of chance.

Nature is ever more ready to assist the doer of good than the evil-doer, and if men and women would but spend a thousandth part of the time, trouble, and wealth, with a view to the improvement of the generations yet to come, that they do to encumber their estates, the amount of human suffering would be vastly reduced, and the world would be much healthier and happier than it is. But so long as men and women will set the hectic flush above the ruddy glow of health, consider curve of lip or eyebrow of more importance than mental power or moral worth, prefer length of purse and social rank to the happiness of their children, let sickly sentiment take the place of reason, "wilfully frustrate the noble purposes of their being, and selfishly ignore the laws of hereditary transmission," so long shall the unfit be begotten, hereditary disease flourish, and immense avoidable suffering continue in the world as the sins of the fathers and mothers are visited upon the innocent children.

CHAPTER V.

THIS principle of prepotency in heredity may be said to act in opposition to that of reversion, and so render the building up of new races and varieties possible. Reversion is conservative; it tends to stamp out all new characters and to continue the race as it is, or was, while the principle of prepotency permits the new character to acquire, in time, strength and fixity sufficient to resist the action of reversion, and ensure the transmission of the new character to the next generation.

When speaking on the subject of heredity, many take it as a fact that all characters in the parental economy are equally potent, and, consequently, stand an equal chance of being transmitted to the offspring. But such is certainly not the case. We have already seen that recently acquired characters are not transmitted with the same certainty that those of long standing are. At first, before the new character has become deeply impressed upon the animal organism, there is a very great liability on the part of the offspring to revert to the *status quo ante;* but with each generation this liability to reversion lessens, as the acquired character becomes more firmly fixed. Thus with every transmission a character becomes more firmly fixed; but to secure such repeated transmission of a new character, it is necessary that each generation should live under conditions very slightly removed from those under which the character was originally acquired, and also, that for a time at least there should be no "crossing;" for, with a changed environment, or with a distinct "cross," there will be reversion, and the character will be lost. Hence the character of long descent is the result, not alone of "in and in" breeding, but of this carried on in an environment very similar to that in which the character first made its appearance.

The peculiar mental character of the Hebrew, of which we have spoken, may here be taken as an example. It was, in the first place, the outcome of a certain condition of environment, that is, the peculiar mode of life followed. Then it was deepened and strengthened, given a prepotency, by intermarriage with other individuals who had existed in the same environment and who had also acquired the character; and it has been preserved and fixed in the race by continued intermarriage and by a continuance of the environment which first called it into existence, viz., the mode of life.

So in every individual, or rather in every family, there are certain characters which stand out prominently amongst the myriads of others inherited,

5

and give tone or bent to the whole. Of the innumerable characters inherited by the child, some are recently
acquired by the family, and are but slightly developed
and still more slightly fixed; others, even of long
standing, are fading from want of cultivation, or from
some change in the environment; while others, again,
are prominent or strongly marked, not from long
descent alone, but from this coupled with continued
cultivation, *i.e.*, by intermarriage among persons of
the same variety in an environment favourable to the
continuance of the character. In fact, prepotency is
another name for fixity, and fixity of a character is
only to be secured by repeated transmission; which,
in turn, is only to be gained by "in and in" breeding
in as nearly constant an environment as is attainable.
Breeders of animals use the word "fixity" in this
sense. When they have succeeded by repeated judicious
"crossing" in obtaining a new character, which they
consider valuable, they set about "fixing" it, which
process consists in maintaining the original environment as nearly as possible, and permitting no "cross"
of fresh blood. When by these means the new character has been transmitted through several generations,
the character is said to be "fixed," that is, it has gained
a prepotency, has become a prominent character in the
animal economy, and is little liable to be lost by reversion. But even when this fixity has been gained,
if it is desired to preserve the character, it is necessary
that the environment should not be radically changed
and that "crossing" should be prevented; for although

a well-fixed character may defy for a time a change
of environment, or one, or even two or three "crossings," yet it will assuredly disappear, should either of
these be persisted in. In the "Southdown" sheep,
for example, certain acquired characters have been
thus securely fixed; yet if we "cross" the "Southdown" with some other variety, the peculiar characters of the former will soon be lost; and similarly,
if the animals be transferred to an environment widely
differing from that in which their peculiar characters
were acquired, as, say, from the rich English pastures
to the bleak Welsh mountains, the animals will throw
off the acquired characters and revert to the original,
or acquire new characters altogether.

It may be taken, therefore, that the longer any
character has "run in the family," the more deeply
is it rooted, so to speak, and the more difficult to
eradicate. And the character which has been handed
down through many generations is often so fixed and
dominant, that it will appear again and again in spite
of repeated "crossings" with fresh blood. Hence
the gravity of an hereditary disease is by no means
to be accurately measured by the symptoms presented
in the individual, but rather by the number of generations through which it has passed to reach him; and
the risk of children inheriting such diseased conditions or predisposition as lead to epilepsy, scrofula,
insanity, drunkenness and the like, will increase proportionately with the number of generations through
which the tendency has been handed down. Thus,

the person who has actually been insane may in some cases marry with far less risk to the children than many other persons who have never been so. For example, the man whose grandfather and father have been insane, would stand a much greater chance of begetting children who would become insane, although he had never shown a symptom of mental disorder himself, than the man with a really good family history whose mind had given way under pressure of some extraordinary trial, mental or physical, and who was some time recovered. In the first case the character has gained prepotency from repeated transmission, and even with a perfectly healthy wife, the chance of reversion is not nearly so great as in the second, where the character being recently acquired and having no fixity will probably disappear, more especially if the wife be healthy, and the exceptional state of environment which developed the character be removed.

Fortunately, there are very few diseased conditions in which we can cite instances of prepotency, as these pathological variations, when continued, soon reach the necessary fatal type and put an end to the family. For this reason hereditary disease is seldom sufficiently firmly fixed as to be able to resist the natural tendency to reversion to the healthy type, if opportunity be offered by marriage with the healthy. And for the same reason it is rarely, even in the most deeply tainted families, that we find the pathological family character reproduced in *all* the children. In some

it is almost certain to appear, but it rarely attacks
every child, for the reason that before the predisposition to disease has gained sufficient fixity in the
family to accomplish this, the family is extinct. Thus
the end of insanity regularly transmitted in a family
is, as Morel has shown, sterile idiocy. Inter-marriage
of those in whom the liability to phthisis is extreme,
though often fruitful, seldom enriches posterity. The
children of the deeply scrofulous are mostly carried
off in infancy and childhood, or drag out a miserable
existence, the inmates of idiot and imbecile asylums.
Epilepsy and drunkenness lead to early and violent
deaths, insanity, idiocy, and extinction, while the
instinctive criminal is the unfortunate representative
of a decaying race. In these cases the predisposition to disease has taken such hold upon the organism,
that the opportunity offered for reversion is not
sufficient to induce that desirable change. The individual is brought forth unsuited to his surroundings, and consequently succumbs. That equilibrium
between creature and environment of which we have
spoken has been lost, and a continuance of the family
has become impossible.

Nevertheless, there are some hereditary diseases
which are handed down through a sufficient number
of generations to gain a considerable prepotency. The
first among these is gout; rheumatism is another in
which prepotency is often attained. In some families
gout has gained such fixity, that it appears in almost
every member of the male line, generation after gene-

ration, and this is to be accounted for by the fact that although gout entails considerable suffering upon its victims, it rarely proves fatal until long after the procreative period has been reached. The untainted blood introduced from time to time is rarely sufficient to rid the family of the disease, because the mode of life which first induced the abnormal condition is persisted in. Here we have a good example of the action of a constant environment acting in opposition to the principle of reversion. When the man who has inherited gout marries a member of a healthy family, the *vis medicatrix naturæ* has only half a chance, so to speak, for while the introduction of untainted blood offers opportunity for reversion in the offspring to the healthy type, the man rarely changes his mode of life, that is, the environment favourable to the abnormal character continues, and so the pathological variation is transmitted, mitigated, it may be, by partial reversion in consequence of the "cross," but still sufficiently potent to ensure, with the aid of the constant environment, its reappearance in the next generation. Like every other hereditary disease, gout is a degeneration, and although it often runs long in a family, there can be no doubt that if the environment favourable to it be maintained, it will ultimately attain the necessarily fatal type, and it is certain that to this hereditary degeneration must be attributed the extinction of many branches of our aristocratic and well-to-do families.

But it is not among the rapidly fatal class of hereditary

characters that we must look for the most convincing proof of the theory that a character gains prepotency with age. Here we do at times find cases to support the theory, but it is amongst those less grave characters which, while unmistakably marked, do not so rapidly go to the extinction of the family that we must find our strongest proof, among such characters as hare-lip, cleft-palate, club-foot, squint, cataract, supernumerary fingers or toes, colour-blindness, premature baldness or greyness, deaf-mutism, stammering, plurality of births, the hæmorrhagic diathesis (bleeders), spina bifida, and the like; or, on the other hand, where the character is physiological. Instances of repeated transmission of any or all of the above-mentioned characters can be found everywhere around, and, doubtless, cases will present themselves to the mind of the reader. It is possible such imperfections may appear in a family in which they have never appeared before, but, in such cases, if the individual bearing the character marry one having no tendency to such characters, the acquired imperfection will not appear in the next, or in succeeding generations. In all families in which these abnormal developments are regularly transmitted, inquiry will elicit the fact that the character has appeared regularly in the family as far back as there is any record of the family. A very good case in point is that recorded by Dr. W. C. Grigg.* He says—" I was consulted by a Mrs. M. B., a Wiltshire woman, aged forty-four, who gave me this history

* *Brit. Med. Journal*, 8th March 1890.

of her family: Great grandmother, maternal side, had nine children at three births, triplets each time. Grandmother had seven children, triplets once, twins twice. Her mother had twelve children, once triplets, twice twins, five single births. Her mother's sister had seven children, triplets once, four single births. Her mother and her aunt married two brothers. Her mother had two brothers who married; neither had children. Mrs. M. B. has had sixteen children, triplets twice, ten single births. She has seven girls living. Eldest daughter aged twenty-five, married, has four children, one triplet, one single birth. Second daughter married September 1889, pregnant. She states that she herself is one of a twin, and her mother also. Her family seems to be well known in the village whence she comes as the 'triplet and twin family.' Her maternal great-aunt, aged ninety, single, is still living, who declares that her grandmother told her that *her* grandmother informed her triplets were in the family as far back as any record could be obtained."

Another example of prepotency of a physiological character is the oft-cited case of the reigning family of Austria, in which a markedly peculiar facial character —the Hapsburg lip—has been transmitted with marvellous certainty through a great number of generations, apparently too firmly fixed to be eradicated or even modified by the infusion of fresh blood which occurs with almost each generation. In this case the peculiar character doubtless arose from some accidental circumstances, and by unpremeditated selection became fixed so that it was transmitted again and again, gaining in fixity with each successive transmission.

The strongest proof of all, however, of the fixity or prepotency gained by long descent, is to be found in reversion. When a character is lost in consequence of reversion, it is nearly always a more or less recently acquired one, while that which appears in its place is invariably an old family one. The normal type, which reversion does its best to reproduce, is made up of characters which have run in the family for ages. Acquired characters have accumulated in the course of a few generations, and the old family type seems to be overborne and lost in the new. But the lines of the old family pattern are not erased. They lie deep down in the organism intact, only blurred or hidden by the recent characters overlying them, and ready to appear, clear and distinct, when these latter from any cause are brushed aside.

But it is unnecessary to labour the point. If we accept evolution, I think we must accept the theory of prepotency increasing with age as a part of that doctrine. Were it not possible for a character to be thus fixed by repeated transmission, so as to defy the natural tendency to reversion to the original, evolution would be an impossibility, and all that has been built upon it must tumble to the ground.

Thus we see that evolution and heredity go hand in hand, they work together, they are inseparable. Evolution modifies the individual and suits him to his surroundings, his mode of life, and heredity perpetuates the modification in his descendants. Were it not so, every change in the individual, brought about by education, training, and mode of life, must of necessity cease to exist with the life of the modified individual, and without evolution, heredity could only reproduce the one changeless type, and all nature would be at a standstill. Ribot says:—" These modifications, as they accumulate and in course of time become organic, make new modifications possible in the succession of generations. Thus heredity becomes in a manner a creative power."

CHAPTER VI.

THE LAWS OF HEREDITY.

WE will now very briefly consider the so-called laws of heredity. These laws are not based on any very scientific foundation, but they are nevertheless most useful when we leave the broad theory and come down to the very interesting study of individual facts.

The following five "laws" are generally given as including all the phenomena of descent.

      I. Direct Heredity.
      II. Reversional Heredity or Atavism.
      III. Collateral or Indirect Heredity.
      IV. Initial Heredity.
      V. Heredity of Influence.

This is not Ribot's, or the usual classification. He omits initial heredity altogether, although it is of the utmost importance, and has perhaps a greater influence upon the child than any other form of heredity excepting only the direct, of which it is really a form.

Let us now glance at each of these rules or laws, and see what part of the great law of heredity each includes.

I. By DIRECT HEREDITY is meant transmission direct from parent to child, that is, where there is least interference with heredity pure and simple, and the child is a compound of its parents. Prepotency may, and in fact generally does come into play here, but there is a marked absence of variation and reversion.

This direct heredity is generally split up into two sub-divisions, thus :—

1. *Where the child resembles each of the parents equally* in its moral, mental, and physical characters— in fact, where the child is an exact mean of its parents. But, as we have already seen, this result, which would be a realisation of the ideal law, must be of extreme rarity, or, if we wish to be scientifically correct, an utter impossibility, for to ensure an exact mean in the child would entail a perfectly equal blending of the parental characters, which prepotency can seldom, if ever, allow, together with a constant environment which can never be obtained. We do occasionally meet with a case where the child appears to be as nearly as possible a mean of its parents, but even in such cases it is never difficult to discover that the mean is far from perfect or exact. For this reason this form of direct heredity needs no consideration at our hands.

2. *Where the child, being a compound of its parents, resembles one parent more strongly than the other.* In most cases of direct heredity the child resembles some one of the parents much more strongly than the other, and this is only what we would expect from what we have already learnt of prepotency. But although the

child may show little or no trace of resemblance to one of its parents, the influence of that parent must not be taken as absent, for the characters peculiar to that parent may be only lying latent in the child, ready to appear in the next, or some more remote, generation. If any one will look around among his relatives and friends he will have little difficulty in discovering cases in proof of this, such, for example, as where a son, who apparently in no way resembles his mother, begets daughters in whom the peculiar characters of their grandmother, although absent in their father, are reproduced with striking truthfulness. But this will be fully considered under Reversional Heredity.

In direct heredity there is, then, in nearly every case, a preponderance of resemblance to one or other of the parents, and this preponderance runs in two ways. 1. Direct, that is, from father to son and mother to daughter; and 2. Diagonally, from father to daughter and from mother to son. Here again the reader will find little difficulty in discovering families which will act as illustrations, for it is a matter of common remark that in some families the sons resemble closely the father and the daughters the mother, while in others the sons have, as a rule, a peculiar resemblance to the mother and the daughters to the father. This will be more easily followed in cases in which the peculiar family character takes the form of some gross variation, as, say, epilepsy or scrofula. In such cases, if the heredity be direct and the father be the parent bearing the taint, the sons will be epileptic or scrofulous, while the daughters may escape; and if the mother be the affected one, while the sons escape, the daughters will be epileptic or scrofulous as the case may be. But if the heredity be of the diagonal order the epilepsy of the grandfather will descend to the mother and through her to the sons, and so on, appearing in different sexes in each succeeding generation.

Of these two the direct is the mode of transmission of parental characters on the whole the more commonly met with, the sons in most families " favouring " the father, and the daughters the mother; and this is what we might expect. But there are some family characters which are said to be much more commonly transmitted by the diagonal than the direct. Among such characters are physical deformities and other imperfections of development, as deaf-mutism, hare-lip, squint, club-foot, supernumerary digits and the like. But although this is generally accepted, there is very little evidence to support it, and I am inclined to doubt whether even in these particular cases the diagonal is the more common mode of transmission.

In support of this view, I would remark that such structural peculiarities as have been mentioned above, are much more likely to prove a bar to marriage in the female than in the male, and as these deformities are as frequently—if not more frequently—met with in the male, therefore they must in a majority of cases be conveyed by direct heredity from the father to the sons. Besides, such peculiarities being a very slight bar to marriage in the male, they are consequently more likely to be handed down through many generations in the male line than in the female, and must thus by repeated transmission gain a prepotency in the male line.

A good example of structural peculiarity transmitted generation after generation in the same sex, or directly, is the facial peculiarity seen in the present reigning family of Austria, "the Hapsburg lip" of which we have already spoken, which has been handed down along the male line for a considerable number of generations, seldom appearing in the female members of the family, and apparently but slightly influenced by the new female blood introduced with almost every generation. A still more peculiar case was that of Edward Lambert, "the human porcupine," as he was called; this man's skin was covered by warty projections which were periodically moulted. He had six sons and two grandsons similarly affected, while the females of the family escaped; the two grandsons mentioned having seven sisters who were free from the peculiarity.

To sum up, then, direct heredity is where the child takes its nature or constitution from its parents; where no prominent character in the child is not to be found in one or other parent. In these cases there is almost invariably a preponderance of resemblance to one or other of the parents, and this preponderance may run either direct, that is, in the same sex from father to son, and mother to daughter, or diagonally—from father to daughter, and from mother to son. Of these, the former—direct—is decidedly the more common in

insanity, epilepsy, scrofula, and gout, while the latter—
diagonal—is said to be most frequently observed in
cases of structural peculiarity.

II. REVERSIONAL HEREDITY or ATAVISM. "This is
a term used to denote cases in which a child, instead
of resembling its immediate parents, resembles one of
his grandparents or still remoter ancestor, or even
some distant member of a collateral branch of the
family" (Lucas). This is a very common form of
heredity. To recognise some peculiar character in the
grandchild which is absent in the parent, yet strongly
marked in the grandparent, may be said to be an
almost everyday occurrence. In some diseases—
pathological variations—this mode of transmission is
so regularly followed that these diseases have come
to be looked upon as only attacking every other generation. Gout thus frequently attacks only alternate
generations, and there are several other diseases which
at times follow the same rule; therefore it should be
understood that the absence of a "family disease" in
one generation is no evidence that the taint has been
shaken off and got rid of, and will not appear in the
next generation. Sir William Aitken states his opinion
that "a family history extending over less than *three*
generations is almost worthless, and may be misleading."

This reversional form of heredity is to be explained
in this way :—Where the peculiar character transmitted
belongs to what might be called the physiological class,
it may be taken that its reappearance in the family is
the result of the action of the natural tendency to

reversion to the original family type; and where the
character is of the pathological order, as in insanity,
gout, idiocy, and the like, it may be taken that the
character, or tendency thereto, is present in every
generation, but in some remains latent all through life,
either because it has been too far mitigated by the
infusion of the untainted blood of the other parent, or
because it has not received some necessary fillip to act
as a starting-point or exciting cause.

Of this latency of characters we have many examples.
The most commonly cited is gout, but there is a much
more remarkable instance to be found in families showing the hæmorrhagic diathesis—commonly known as
" bleeders." Here the peculiar morbid condition is
purely hereditary, and although it has been rarely,
if ever, seen in the female, it is regularly transmitted
through the females to the males of the next generation. It may even be transmitted through two or
three generations of females, to reappear in the males
so soon as that sex appears in the family. Dr. Wickham Legg * and Dr. Finlayson † have studied this subject, and have published family trees illustrating this
strange fact. In explanation of this " sexual atavism,"
if it may be called so, of the hæmorrhagic diathesis, I
would submit that the morbid condition present in
these cases was almost from the first necessarily fatal
to the females, it being impossible for a woman of this
diathesis to pass safely through parturition, or even

* St. Bartholomew's Hospital Reports, 1881.
† *Glasgow Medical Journal*, July 1882.

those functions which must precede maternity. For this reason only the offspring of those females in whom the character was latent ever came into being, and so in time this came to be the type of the degeneration, being the only type which it was possible to propagate. Only through the female in whom the character was absent, or latent, could the variety be continued.

Atavism or Reversional Heredity, although frequently met with in the transmission of physical peculiarities, is, perhaps, more often met with where the character is of a moral or mental nature. In what is known as moral insanity it is often met with, while in the ordinary intellectual mental disorders it is very common, the offspring of him or her who is of neurotic family and who has actually been insane frequently escaping the insane temperament, or even occasionally developing high talents, approaching genius perhaps, while in the next generation the children, instead of inheriting the high mental characters of the parent, revert to the insane type once more. I would here remark that the genius springing thus at intervals from the insane stock is not of the highest description, and notwithstanding the opinion of so weighty an authority as Dr. Maudsley, who fears that in forbidding the marriage of those of the insane temperament, we would to a certain extent be stamping out genius in the race, I would venture to express the opinion that the chances of the insane parent enriching the world by begetting the genius, are not sufficiently good to justify the insane in hazarding the experiment.

III. INDIRECT OR COLLATERAL HEREDITY is said to occur when a child resembles mentally, morally, or physically, some relative out of the direct line of its descent, as an uncle or aunt. This is in reality not a distinct form or mode of transmission, but only a modification of the atavism we had under the last head, and, like it, is to be explained by reversion. That is to say, the parents in the direct line are variations from the original family stock, while the collateral descendants have followed it more nearly ; and the children, in displaying the characters common to such collateral relatives, and absent in their parent, are merely showing a reversion to that family type from which their parent had varied.

IV. INITIAL HEREDITY. This depends upon the temporary mood or condition, good or evil, fortunate or unfortunate, of the parents when they become such. This form of heredity is not given by Ribot, but its existence is perfectly well established, and it has, perhaps, more influence in ordaining what the child shall be than any other heredity except the direct only. In fact, it is a form of direct heredity, and, save that it deals only with temporary conditions, and at a particular time, it might be included under the first law.

An American writer says :—" A good initial heredity may produce virtue in the descendants by predisposition merely from a temporarily ennobled nature although there was a general vice in the parents, and so a bad direct heredity. If you are in a lofty mood, Providence is on your side ; but when a drunkard on the one hand, or when, on the other, a man generally temperate, but in a temporary debauch, places himself under the power of this law of heredity, the specific or initial principle acts just as surely to produce an inheritance of evil, as it does in the opposite case to produce an inheritance of good." This may be a little too poetically expressed, still it is very near the truth. Whether a temporary elevated or ennobled condition in the parent can override a fixed family viciousness we are not in a position to assert positively, but that a temporary depraved condition can override the best family predisposition we know for a fact, and why this law should not act for good as well as for evil in the race we fail to see. As regards drunkenness, Dr. Maudsley says :—" Here, as elsewhere in nature, like produces like; and the parent who makes himself a temporary lunatic or idiot by his degrading vice, propagates his kind in procreation, and entails on his children the curse of the most hopeless fate."

That drunkenness and other vicious temporary conditions in the parent, when he becomes such, have a powerful influence for evil upon the child begotten, has long been known amongst the people, and of its truth some proofs have recently been collected. Cases are by no means rare in which a temporarily drunken parent has begotten an idiot child. Indeed, several observers have collected statistics which go to prove that the vast majority of idiots and imbeciles, who are not the result of a family degeneration, are the children of drunken and otherwise vicious parents, and it would not be too much to infer that much of the mental and moral obliquity and degradation met with in the poorer classes, from which springs the instinctive criminal, has its origin in vicious initial heredity. Let us hope, then, that this law of Nature is as active for virtue as for vice, and take it that not a little of that which makes human nature lovely, is the outcome of a pure and ennobled nature in the parent when he becomes such, for nothing but good can arise from the teaching of such doctrine.

Laurence Sterne shows a deep insight into the ways of Nature when, in the opening lines of "Tristram Shandy," he says:—"I wish either my father or my mother, or, indeed, both of them, as they were in duty both equally bound to it, had minded what they were about when they begot me; had they duly considered how much depended upon what they were then doing, that not only the production of a rational being was concerned in it, but that possibly the happy formation and temperament of his body, perhaps his genius, and the very cast of his mind, and, for aught they knew to the contrary, even the fortunes of his whole house might take their turn from the humours and dispositions that were then uppermost. Had they duly weighed and considered all this, and proceeded accordingly, I am verily persuaded I should have made quite a different figure, on the whole, from that in which the reader is likely to see me. Believe me, good folk, this is not so inconsiderable a thing as many of you think it."

V. HEREDITY OF INFLUENCE. This is seen when the child of a second marriage resembles the husband of a previous marriage; as, for example, when a white woman has had children by a first husband who was a negro, and afterwards has children by a second husband, who is of white blood, and the children by this second marriage show distinct evidence of the presence of characters peculiar to the negro. It appears that the male who first impregnates a female, so impresses the organism of that female that the young she bears to other males will more or less "favour" him who first impregnated the mother. This is well known to the breeders of animals, who are most careful not to permit the approach of chance males, as the influence of such a cross will frequently be observable in many following pregnancies by other males. Nor are we without evidence of the working of this law in the human family; but, as far as man is concerned, it is of no importance except to the medical jurist, whom it sometimes aids in deciding paternity.

These, then, may be taken as the so-called laws of heredity, which have been formulated to include most of the phenomena met with in families. They are but arbitrary divisions of the one great law or principle, and may be increased in number, or varied, at will. Some writers go further in their classification of hereditary phenomena, and make a large number of divisions and sub-divisions; but this serves no good purpose, and often only tends to confuse the student. There is, however, one class of cases which, while not deserving inclusion under a separate head, may with advantage be here briefly considered. This is what has been called—

HEREDITY AT CORRESPONDING AGES. Here the peculiar character may have been transmitted directly from the parents, or it may have been the outcome of reversion—in fact, it matters not how it has been acquired, the sole peculiarity to be noticed being the fact that the character makes its appearance at a certain age, and that age is, in the child, the same age as that at which the character had previously appeared in the parent or other ancestor. This peculiarity is most frequently noticed in phthisical families, but it often occurs in cancer, insanity, and other transmitted degenerate conditions. It is a matter of common observation that in some phthisical families the children grow up strong and apparently healthy, but on attaining a certain age the inherited disease, or perhaps I should say the disease, a predisposition to which has been inherited, lights up, and one after another they die off. Austin Flint, in his excellent "Practice of Medicine," when speaking on this point, says:—"This congenital predisposition may remain completely latent until the period of life in which the disease is most apt to be developed; and we sometimes see a whole family of children, one after the other, fall victims to this disease [phthisis], when they severally reach a certain age." Of course our recently acquired knowledge of the character of this disease will modify largely the views once held as to the hereditary transmission of phthisis, but for the present we shall not comment on the above.

In insanity this " heredity at corresponding ages " is frequently met with, as it is also in cancer, gout, and rheumatism. A case in point comes to my mind. It is this:—A father (of whom I could get but little information) was addicted to drink, and became insane at about forty years of age. He had four sons. The eldest became insane at the age of forty-one, the second at the same age also became insane, while the third and fourth sons were in turn certified insane on reaching the age of thirty-eight. They were all, like the father, given to habits of intemperance, and not one of them ever showed any sign of mental improvement after the first mental failure. Each sank from bad to worse, and soon arrived at hopeless dementia. The eldest of these brothers is at present a murderer in Broadmoor criminal asylum, and the other three are, as I have said, hopeless dements in a county asylum.

And now, having considered the courses generally followed by Nature in the transmission of hereditary characters, we will glance at a few of the more grave pathological characters commonly transmitted, and consider what advice should be offered persons bearing such taint when the question of marriage arises.

CHAPTER VII.

HEREDITY IN INSANITY.

"Legislative enactments regarding the inter-marriage of persons tainted by disordered intellect are greatly to be desired; and the concealment of such disorder, with a view to marriage, ought to render marriages null and void which are concluded under such circumstances."—SIR WILLIAM AITKEN.*

INSANITY has been at all times in the world's history the most dreaded of infirmities, and rightly so, for no other diseased condition whatever inflicts so grievous suffering, not only upon its victims, but upon all those nearest and dearest to them. We look upon the disease as cruel which tears the innocent child from its mother's breast; which chills the warm blood and for ever stops the bounding heart of youth, or lays the young father or mother cold in death before the eyes of the terror-stricken children; such disease we look upon as cruel, and with eyes blinded by tears for those we loved, it is sometimes very difficult for us to see in all this the hand of a kind and merciful Providence. But if we contrast the brief suffering of the dear ones we have lost with the lifelong agonies of many of the insane, we must rejoice, and say, in truth, that death

* "The Science and Practice of Medicine," p. 490, vol. ii.

has been "cruel only to be kind," for rather a thousand times the quiet forgetfulness of the tomb than the lifelong battle of the chronic maniac, the imaginary, but no less torturing hell in which the melancholiac exists, or the living death of the dement.

And if it be certain that no other disease causes such terrible suffering and degradation in its victims, it is still more certain that there is no "ill that flesh is heir to" which creates a tithe of the misery and distress amongst the relatives and dependents of those afflicted. Need I speak of the agonies of the young wife, but yesterday full of life and hope, whose partner has been dragged shrieking from her side, leaving her wedded yet widowed, she and her young children a charge upon the cold world, or perhaps colder friends; or, on the other hand, the terrible position of the husband who has learnt to look with terror upon the approach of what should be a time of family rejoicing, and who must curse the day he became a father, when he thinks of the future of his children. Well might the ancients imagine such things could but come from the devil.

According to the last report of the Commissioners in Lunacy (June 1891) there were no less than 86,795 "lunatics, idiots, and persons of unsound mind" in England and Wales. It must be remembered, too, that these figures, while representing the great mass of our insane population, by no means exhaust it, for there are hundreds of senile dements and idiotic and imbecile children, epileptic and otherwise, who, belonging to the

middle classes, are kept at home among their relatives
and friends, and so never come within the knowledge
of the Board at Whitehall. However, taking the
Commissioners' figures alone, we may say that there
is one insane person to every 300 of the population,
which is a statement sufficiently startling in itself.

There is a belief abroad that insanity is on the
increase among the people of these countries, and certainly the figures set forth year after year by the
Commissioners in Lunacy go far in support of such
belief. If we take the totals at decennial periods we
find the insane population of England and Wales
increased alarmingly, thus :—

Total insane on January	1859	.	.	.	36,762
,,	,,	1869	.	.	.	53,177
,,	,,	1879	.	.	.	69,885
,,	,,	1889	.	.	.	84,340

Here we have a steady increase in the insane population of England and Wales at a rate of over 1500
a year. Nor is this increase to be accounted for by
increase in the general population, for the Commissioners' own figures show that the proportion of
insane to every 10,000 of the population was on

January	1859	.	.	.	18.67 to the 10,000
,,	1869	.	.	.	23.93 ,,
,,	1879	.	.	.	27.55 ,,
,,	1889	.	.	.	29.07 ,,

Many men learned in lunacy and well able to
form sound opinions on the subject, have reluctantly
admitted, that upon other grounds than an increased

liability to insanity amongst the people it is impossible satisfactorily to account for this alarming and steady increase in our insane; while some others, equally familiar with the subject, have endeavoured with praiseworthy zeal to prove that this increase in the numbers of certified lunatics in proportion to the population is entirely due to the fact that persons suffering from mild forms of mental derangement, such as would have passed almost unnoticed a few years back, are now admitted to asylums there to be taken care of; and further, that the inmates of our asylums are now-a-days so well cared for that their lives are considerably prolonged and they go to swell the numbers of the registered insane years after they would, under the older modes of treatment, have passed over to the majority.

In this way do some writers endeavour to account for the terrible accumulation of lunatics which has been going on for the past quarter of a century or more; and certainly the fact that most of the village fools and eccentric wanderers so common in the last generation have disappeared from their usual haunts, is proof that in some degree the great increase in the population of our asylums is in this way to be explained. But that it can be wholly put down to this ingathering and preservation of the " weak ones " has by no means been satisfactorily proven, and I must admit that I go with those who believe mental disease to be on the increase in these countries.

In support of this belief I would point out that if

we admit, as I hold we must, hereditary taint to be a predisposing cause of insanity, we can come to no other conclusion. There is no class of diseases so surely transmitted from parent to child as the nervous—upon this point the whole medical profession is agreed ; and that our present laws for the management of the insane and those who have been insane tend directly to spread insanity, epilepsy, and allied diseases amongst our people, a moment's consideration will prove conclusively. Take, for instance, the case of a young man who, in consequence of inherited nervous instability, becomes insane. He is treated in an asylum, and as soon as he recovers from the acute attack he is discharged, however bad his family history may be. Being naturally impulsive and emotional, and having but slight control over his passions, he not infrequently marries early, perhaps a very short time after his discharge from the asylum, and when he returns to the asylum—as he is almost certain to do—he is probably the father of two or three children. Again he recovers, and again he returns home to beget a tainted race. Ultimately, in all probability, this man returns to the asylum to remain there, but before that stage in his downward course is reached he has possibly left a large family behind, some of whom will most likely join him in the asylum before he dies. No large asylum is without scores of such cases ; they make up a large part of the moving population of such institutions. At present one case comes to my recollection. It is that mentioned in the last chapter, where a neurotic father had four sons, each of whom on attaining thirty-eight to forty years of age became insane. These four men must be kept for the remainder of their lives at the public expense. But that is not the worst. Three of these men married, and before they had become sufficiently insane to be relegated to an asylum had become the fathers of thirty-four children. Nature fought against this propagation of the unfit, and permitted only thirteen of the thirty-four to reach maturity. One of these has since dropped dead leaving no issue, but twelve are still left as a legacy to the coming generation of ratepayers. Men like these, or those others who also form a large class, who beget families in the intervals between attacks of mania, melancholia, or epileptic excitement, must increase the insane population, and the system which permits such propagation must not be surprised when it is called upon to build new asylums or add block after block to the old.

Or, again, take the case of a woman cursed with a bad inheritance; she marries, becomes pregnant, and, unable to bear the strain thus thrown upon the system, her mind gives way and for a time she becomes an inmate of some asylum. In the majority of cases she too recovers for a time, and goes out into the world to bring forth perhaps a large family loaded with a double allowance of original sin. Every asylum medical officer is only too familiar with such cases. I can call to memory a score such at the present time, women who return to the asylum time after time, each visit in many cases following or preceding the birth of an unfortunate child.

Now, this procedure can have but one result, and that is, the cultivation and increase of insanity and other nervous diseases and degenerations—as epilepsy, chorea, deaf-mutism, suicide, hysteria, idiocy, and the like, and Sir William Aitken is certainly justified in asking that such tainted persons should not be permitted to contaminate the race by propagating their like.

There is another matter which requires explanation before we can admit that insanity is not on the increase, and that is suicide. If the supposition that the increase in the number of certified insane is entirely due to the gathering together of nearly all the insane in the asylums be true, then it follows that the proportion of insane outside asylums should be proportionately diminished, and consequently suicide, which we may take in the majority of instances to be the outcome of mental disorder, should be much less frequent than it was before the insane were so carefully weeded from the general population. If the theory of those who say that insanity is not on the increase be sound, deaths from suicide amongst those outside asylums should diminish. But what is the fact? On reference to the Registrar General's reports, we find that deaths from suicide are increasing year by year, much as the certified insane are. The number of deaths from suicide recorded in 1864 was 1340; in 1870, 1554; in 1875, 1601; in 1880, 1979; in 1885, 2007; and in 1888 it had increased to 2308. Nor is this increase apparent only, for while the proportion

of deaths from suicide was in 1864 only 64 to the
million, it had risen in 1888 to 81 to the million;
which is an increase of as nearly as possible 33 per
cent. within less than 25 years. How those who
maintain that insanity is not increasing explain these
figures of the Registrar-General I do not know, as the
matter has not been considered in this connection, so
far as I am aware; but I fail to see how they can
reconcile the fact that suicide, which is an unmistakable sign of what we know as the insane temperament,
is increasing among the people, with their assertion
that the insane have been winnowed from the people
to an extent hitherto unknown, and that in the mean-
time there has been no increase of insanity. Until
this be satisfactorily explained, I must decline to
believe that insanity is not more prevalent now, when
suicides rank in the Registrar-General's report at 81
to the million, and we have 86,067 certified lunatics
in asylums, than when the certified insane were less
than half and the suicides 33 per cent. under what
they are at present.

The marriage of those deeply tainted with insanity
or predisposition thereto is under any circumstances
to be deplored, but what makes such marriages more
terrible is the fact that in a great many cases the
tainted one is married on the assumption that no such
bar exists, and the unfortunate partner only discovers
when it is too late how cruelly he or she has been
treated by the one above all others implicitly trusted.
Concealment of such family blight under the circum-

stances is a moral wrong, and it must quickly be made a legal one. The Legislature must step forward and say that deception of this kind, deliberately practised, as it too frequently is, with a view to marriage, shall be sufficient ground for a nullification of the marriage contract. If a bill were at present brought forward so to alter the law it would receive the united support of the scientific and legal schools of thought.

Unhappily, however, such deception is not always necessary; for, strange as it may appear to some, we know it to be a fact that many, well knowing the family history of the tainted one, disregard it. This disregard arises from various causes. In some it is the outcome of ignorance, and here from education we may anticipate good results. In some others it arises from gross carelessness, which is nothing short of criminal. But in the great majority of those cases in which the laws of Nature are disregarded, and more especially amongst the educated classes, the offenders are guided solely by sordid and selfish motives, such as social elevation and love of wealth. In all civilised countries, in the highest families—not excepting royalty itself—we find men and women, for their own personal aggrandisement, deliberately, we might almost say with malice aforethought, entering into marriages which can only end in disaster to the luckless children. Some puling sentimentalists, who cannot plead ignorance, assert that they are led by love's legendary single hair, but such people are not to be believed. A bowstring would not drag a vigorous

7

and right-minded man or woman to such a fate. These people are actuated by pure selfishness, as is another class, the quasi-religious, who throw all responsibility upon Providence; canting blockheads who forget that God helps those who help themselves, and who refuse to understand that Providence, having established benign laws for the government of His creatures, will not stultify Himself by staying those laws in answer to the whine of those who have wittingly disregarded and violated them. In all these cases, high and low, selfishness pure and simple is the motive power, and the strong arm of the law should be invoked to prevent, so far as is possible, such selfishness saddling the community with a helpless, worthless offspring.

Still, in the great majority of cases among the middle classes deception is practised. Rightly or wrongly, people look upon insanity or epilepsy as a stigma upon the family, and use every endeavour to preserve its existence unknown to the world. The answer to the question, " Is there any insanity in your family ? " even from those whose word upon any other question might be implicitly relied on, is often not worth the breath which gives it utterance. Only the other day I asked a lady, whose daughter was insane, whether any other member of the family had ever suffered from any mental or nervous disease of any kind, and she hastened to assure me that such a thing was unknown to the family. Having had a different family-history from the insane daughter's unfortunate husband, I was rather taken aback at the reply, but

guessing upon which side the error lay, I ventured
the further question, " Did not another daughter of
yours commit suicide ? " To which the lady replied
without a blush, " Oh yes, I had forgotten that."
On this point Maudsley says :—" The more exact and
scrupulous the researches made, the more distinctly
is displayed the influence of hereditary taint in the
production of insanity. It is unfortunately impossible
to get exact or accurate information on this subject.
So strong is the foolish feeling of disgrace attaching
to the occurrence of insanity in a family, that people
not apt usually to say what is not true, will disclaim
or deny most earnestly the existence of any hereditary
taint, when all the time the indications of it are
most positive ; yes, when its existence is well known,
and they must know that it is well known. To elicit
an acknowledgment of the truth in some of these
cases, would be as difficult a task as to elicit from an
erring woman a confession of her single frailty."

And yet, in spite of this hard lying on the part of
relatives, alienists have been able to trace distinct
hereditary taint in a large proportion of the cases
coming under their observation. Moreau put his percentage as high as 90. Burrows said 85, Holst 69,
Jassen 65, Michéa 50 to 75, Thurnam 51, Webster
32, Needham 31, Guislain 30, Maudsley 28, and
Esquirol 25.

These figures vary widely. They vary with the
amount of prevarication and untruth practised by the
relatives of the insane, and it is to be feared that

until human nature becomes something different from what we know it to-day, or until families are compelled by law to keep some kind of family record, little more than we at present know on this most important subject will be learnt from statistics. From education, the modern cure for all ills, we can expect nothing, for we find that in the upper classes, where education should be most advanced, truth upon this one point at least is less plentiful than among the ignorant.

The Commissioners' summary of the whole number of persons certified as insane in 1887, shows that in spite of error, accidental and premeditated, close on a fourth—23 per cent.—were by heredity predisposed to insanity, while of the total admissions for the ten years 1878 to 1887 inclusive, in 20.5 per cent. inherited taint was admitted.

Now, in view of this conclusive evidence of the hereditary transmissibility of so terrible a disease, a disease whose ravages in society scientific men and economists alike deplore, and whose increase under the existing state of things medical science is unable to stay, I think the time has arrived when something should be done to limit its propagation, either by teaching the people, or, as a last resort, calling in the aid of the Legislature. Ultimately, I fear, this latter course must be adopted, for the reason that many of those of the insane temperament are so ill-balanced, emotional, and impulsive, that they are at best only semi-responsible, and the teachings of science, however convincing to the thoughtful, can never have any great weight with them. Besides, they are backed up in their resolves by their relatives, who are about the worst counsellors they could have. Every one who has had much to do with the insane and their relatives has noticed the mental peculiarities so often exhibited by the latter. Very frequently their minds are crippled and deformed. They are obstinate, passionate, wilful and suspicious, the higher intelligence being often replaced by a deep low cunning, while the moral side of their nature is equally poorly developed. Weak themselves, and with such counsellors, what can we expect?

I have seen a man, whose mother was an imbecile, whose sister was an idiot, and who was little better himself, come to visit his wife and wife's sister— whose mother had also been insane—who were confined as lunatics in the same asylum in which his idiot sister resided, and I have watched this creature, the father of three children! laugh gleefully at the antics of his relatives in the visiting-room. There are thousands of such ill-developed men and women in the country, creatures whom we cannot hope to guide otherwise than by force. Education is all very well in its place, and it must have a most beneficial effect amongst those who have sufficient mental development to appreciate the evil under which their families labour, and who have sufficient strength of will to enable them to choose the good rather than the evil. But with those like the man above mentioned it is useless to plead, and only coercion will keep

them in the right path. They attend upon the calls
of their instincts and passions as does the unreasoning
beast, and not even an angel from heaven could hope
by moral suasion to induce them to curb a single
appetite or in any way mortify the flesh.

Of course the old cry of "interference with the
freedom of the subject" will rise like a spectre to bar
the path of legislation; but this ghost has been laid
before and will be again. These wretched creatures,
far down in the scale of degeneration, with just sufficient intelligence to keep them from outraging the
usages of society; who continue their kind so long as
nature permits, to the detriment of the race; who
create nothing, add nothing to the commonwealth,
but are, instead, a charge upon the community;—these
have no more right to claim freedom of action as to
procreation, than has the leper to mingle with the
populace.

All men and women who have been insane once and
have a bad family history, those who have been twice
insane, even if the family history be good, and all who
are confirmed epileptics or drunkards, should be prevented by the state from becoming parents, for they
have no greater right to carry suffering and contamination amongst the people, and throw expense upon
the state, than has the person suffering from small-
pox to do so by travelling in a public conveyance.
As with the victim of the small-pox, it is their misfortune more than their fault, but of this society can
take no notice. The unfortunate few must always

snffer for the benefit of the many. It is the duty of the state to see that such unfortunates are tended and cared for, and that their lives are made, so far as is possible, bright and cheerful. But that they should be permitted to hand down their disease to innocent children, any more than the sick one should give his small-pox to his neighbours, is unfair to society and to the race.

CHAPTER VIII.

MARRIAGE AND INSANITY.

"If we are seriously minded to check the increase or lessen the production of insanity, it would be necessary to begin further back, and to lay down rules to prevent the propagation of a disease which is one of the most hereditary of diseases."—MAUDSLEY.[*]

THIS is a subject upon which it is difficult to speak, knowing, as we do, that every word spoken must crush the fondest hope of some unfortunate fellow-creature. However, a duty should not be shirked simply because it is unpleasant. Too long has sentiment been allowed to rule our better judgment in this matter, with what result we see, and the sooner we break new ground the better. Discussion must cause pain to many, but silence on the subject would be even more cruel; for, as it has been in the past it would be in the future, the cause of much suffering, sin, and death, which otherwise might never appear on the face of the earth.

First, then, I would advise that every person who knows the family to which he belongs to have the bar-sinister of insanity upon its escutcheon, before daring to put himself, or herself, in the way of

[*] "Responsibility in Mental Disease," p. 275.

becoming a parent, should carefully examine the family tree, get to understand exactly his own position, then lay the whole case candidly and honestly before the physician, and abide by his decision.

Further, I would advise that all persons who contemplate matrimony, all to whom attentions and overtures are being made with a view to marriage, should look upon a mutual exchange of confidences upon this matter of hereditary or family disease as absolutely essential, and that, too, at an early period of the intimacy, before the affections have become deeply engaged. Too often knowledge of the existence of the family skeleton, when given at all, is only given when matters have gone so far that only those of strong will find it possible to give up the loved one because of an evil so distant and shadowy as this family taint appears in the eyes of the lover.

In the majority of families in which mental disease is transmitted, it appears in only one, two, or three members of each generation; but those who do not become insane often bear the taint as surely to the next generation as do those who have actually been insane. In rare cases, insanity or some allied nervous disease attacks every member of a family, but before this state of things is reached the family is very deeply saturated with disease, and is fast approaching extinction. Cases are on record in which as many as eleven members of a family have been insane. I myself have met with a case where, in a family of nine children, six died within the first year of life, every one torn by

convulsions, while two of the remaining three were
"weak in their minds," and the third was a jabbering
idiot, the inmate of an asylum.   Such lamentable
cases are generally the result of marriages in which
both parents belong to the neurotic or insane type.
The deeper the taint the less likely are the children
to escape it, and nothing so certainly tends to deepen
the taint as "in and in" breeding.   The children of
a person come of an insane stock will be infinitely
more likely to escape the family blight if that person
marry a healthy person, than if he marry one come
of a family like his own, and every effort should be
made to impress upon the public mind the danger of
inter-marriage amongst neurotic families.

The person, man or woman, who has had an epileptic,
or choreic, or imbecile brother or sister, an insane uncle,
aunt, or parent, or even grandparent, should never for
a moment permit himself to look upon a member of
any neurotic family—that is, one in which insanity,
epilepsy, habitual drunkenness, suicide, or imbecility
has at any time appeared—as a probable, or even
possible, partner in marriage; for although the disease
has appeared in but one or two members of the family
it shows that the tendency is there, and the chances of
the children not inheriting disease from such a union
will be very slight.   Those of neurotic family should
never forget that the safety of their children wholly
depends upon their choice of partners.   If they marry
one not of their own type the tendency to nervous
disease—if not too deeply marked in themselves—may

be totally lost in the children from the action of the
vis medicatrix naturæ; whereas if they marry into a
family like their own, there will be little chance of
reversion to the healthy type, most likely the peculiar
conformation or temperament which predisposes to
insanity will be accentuated in the children, and probably idiocy, imbecility, epilepsy, chorea, drunkenness,
crime, suicide, or insanity will make its appearance in
the degenerate offspring, as in some of the families
whose histories are given in these pages.

All these diseases, together with neuralgia, hysteria,
cancer, and the like, are allied, and, following some law
at present unknown to us, replace each other in
successive generations, and in different individuals of
the same generation in a manner at present inexplicable (*vide* p. 49). Thus the son of an insane parent
may be a confirmed drunkard, and he in turn may
beget a family one member of which may inherit his
father's vice, while another may be epileptic, another
idiotic, and yet another who, perhaps after giving early
promise of superior intellectual attainment, will become
insane. But although these allied diseased conditions
do exchange places in unaccountable manner in neurotic
families, we are not by any means without examples
of the transmission of the same form of mental disease
through several generations, or even of the same mental
disease appearing at almost exactly the same age and
under like circumstances.

Esquirol was of opinion that as a rule the same form
of mental disease was transmitted, and this opinion was

confirmed by Moreau, who says :—"It is rare that the form the malady assumes does not present the most striking resemblance, sometimes even a true identity." During ten years' experience amongst the insane I may say I have met with the same form of mental disorder repeated in different members of the same family sufficiently often to induce me to agree to a great extent with this dictum of Moreau. But although every form of insanity tends thus to be transmitted unchanged in the family, all forms are not equally stable, some being much more commonly transmitted unaltered than others. Of all forms of mental disease by far the most certainly transmitted unaltered is the propensity to suicide ; the other forms in order of frequency following thus—dipsomania, melancholia, monomanias, mania, imbecility.

Dr. Stewart of the Crichton Institution, after a study of 901 cases of mental disease, gave the proportion of hereditary cases in the different forms of insanity thus : Melancholia, 57.7 ; dipsomania, 63.4 ; mania, 51.0 ; monomania, 49.0 ; moral insanity, 50.0 ; general paralysis, 47.6 ; and idiocy and imbecility, 36.0 per cent.

A striking example of transmission of the suicidal propensity is given by Dr. Hammond of New York,* the case being the more remarkable from the fact that exactly the same means were adopted to destroy life and at about the same age in each of the three generations : " A gentleman well-to-do in the world, but with a slight hereditary tendency to insanity, killed himself in the thirty-fifth year of his age by cutting his

* " Insanity and its Medical Relations."

throat while in a warm bath. No cause could be assigned for the act. He had two sons and a daughter —all under age at the time of his death. The family separated, the daughter marrying. On arriving at the age of thirty-five the eldest son cut his throat while in a warm bath, but was rescued ere life was extinct. At about the same age the second son succeeded in killing himself in the same way. The daughter in her thirty-fourth year was found dead in a bath-tub with her throat cut. Her son at the age of twenty-seven attempted to kill himself by cutting his throat while in a bath at his hotel in Paris, but did not succeed. Subsequently at the age of thirty he made a similar unsuccessful attempt, but was again saved. A year afterwards he was found in his bath by his servant with his throat cut from ear to ear."

Voltaire records and comments upon a like case in these words:—"I have almost with my own eyes seen a suicide whose case deserves the attention of physicians. A man of serious turn of mind, of mature age, and of irreproachable conduct, free from strong passions, and above want, killed himself on the 17th October 1769, and left a written explanation of his act, addressed to the council of the city in which he was born. This it was thought best not to publish, for fear of encouraging others to quit a life of which so much evil is spoken. In all this there was nothing astonishing—such cases are met with every day ; but the sequel is more remarkable. His father and his brother had each committed suicide at the same age as

himself. What hidden disposition of the organs, what
sympathy, what combination of physical laws caused
the father and his two children to perish by their own
hands, by the same method, and at the same age?"*

Falret gives the case of a family in which the grandmother, mother, and grandchildren were the subjects
of suicidal melancholia, and records the history of
another family thus:—The father was of a taciturn
disposition; he had six children, five boys and a girl.
The eldest, aged forty, precipitated himself from the
third storey, without any motive; the second in age
strangled himself at thirty-five; the third threw himself from a window in attempting to fly; the fourth
shot himself with a pistol; and, lastly, a cousin jumped
into a river from a trifling cause.† Scores of such
cases might be quoted; they are familiar to every
asylum physician. But I need not load these pages
with such melancholy records, enough has been already
given to make clear to the reader with what fatal
certainty this tendency towards self-destruction is
handed down from parent to child. There is one
other point, however, upon which I would like to say
a word, and that is, suicide amongst children. Fifty
years ago suicide of children of tender years, which
has of late become so painfully common, was almost
unknown. Some put this down to an earlier development of the mental powers in consequence of forced
education, that is to say, the period of reasoning

* "Dictionnaire Philosophique."
† Bucknill Tuke's "Psychological Medicine."

discretion is arrived at at a much earlier age than
formerly. To this I have two objections—first, that
suicide was almost unknown amongst the children of
the classes of fifty years ago, although they were
highly educated, and frequently at an early age; and
second, that this argument must be based upon the
assumption that suicide is the outcome of healthy
reasoning, which I think few will admit. Education,
forced at too early an age, has doubtless something
to do with this lamentable increase of child suicide,
inasmuch as it conduces to the building up of those
disordered nervous conditions from which is evolved
the insane temperament; or it may act as an exciting
cause in an ill-balanced and ill-developed mind, but
beyond this it is seldom responsible for these child
suicides. The real cause is inherited taint, just as it
is in the adult. Some hereditary defect will be found
in all such cases, if a sufficiently careful search be
made. "If the child's family history be inquired into,
it will usually be found that a line of suicide, or of
melancholic depression with suicidal tendency, runs
through it. So it comes to pass that a slight cause of
vexation is sufficient to strike and make vibrate the
fundamental life-sick note of its nature." *

The next most regularly transmitted diseased nervous
condition is dipsomania. This we will consider later,
and for the present pass on to

Melancholia.—This, which is one of the most painful
forms which mental disease can assume, has long been

* Maudsley in *Fortnightly Review,* May 1886.

noted for the persistency with which it clings to a family, often appearing at about the same age generation after generation, and even the same delusions—as of ultimate condemnation, impending poverty, and the like—appearing again and again. When a person falls into melancholia it is usually set down to ill-health, or to over-work, to real business trouble or domestic affliction. Doubtless in some few cases this may be true, but in all cases hereditary taint should be suspected and searched for, for no form of insanity is so frequently attributable to this cause, except only the tendency to suicide, and the drink-crave. As Esquirol says:—"Melancholiacs are born with a peculiar temperament, which disposes them to melancholy."

This regularly transmitted melancholia which appears in youth and middle life may, or may not, be accompanied by a tendency to suicide. In some melancholiacs the impulse to self-destruction is ever present, in others it only appears with periodic exacerbations of depression, while in another and still more painful class the unholy fear of horrors to be experienced in the next world makes the sufferer cling to his wretched life with the tenacity of despair. Doubtless many of those cases in which suicide appears in succeeding generations might be put down to inherited melancholia with suicidal tendency, for it is almost impossible to distinguish those cases in which mental depression and weariness of life precede and lead up to the act of self-destruction from those in which the blind impulse to leave the world exists alone.

Often the imaginary troubles of the melancholiac are long borne in silence. Frequently the sufferer lets his unreal woes "like a worm i' the bud" eat at his vitals unknown to the world, and, if not betrayed by his worn and gloomy exterior, having lost hope of relief in this world, seeks it in the next, taking his secret with him.

A good example of the wonderful influence of inherited taint is seen in those painful cases of melancholia which we so frequently meet with among those of atheromatous habit who have passed the meridian of life. In these cases, when the vessels become so loaded with earthy matter as to be impervious to the blood, the surrounding tissue undergoes the usual degenerative changes consequent on starvation. Now, if the patient be of stable nervous temperament, he will, as the nervous degeneration proceeds, sink quietly through his second childishness into the oblivion of dotage; but if, on the other hand, he has inherited the insane diathesis, delusions of persecution, of impending poverty, or of eternal condemnation will arise to make miserable the evening of his life.

These, then—suicide, dipsomania, and melancholia —are the forms of mental disease which are most frequently transmitted unaltered along the family line, but the others (as mania, monomania, moral insanity, propensity to crime and idiocy) may, and sometimes do, appear again and again in families. Lucas quotes Haller, who gives the case of "two noble families in which idiocy had appeared for nearly a century when

8

he wrote, and in which it still appeared in some members of the fourth and fifth generations."

It may be said at once that no family is safe, any member of which has suffered any form of mental disorder, from whatever cause. Of course there are cases where mental disorder follows prolonged intemperate habits, injuries to the head, sunstroke and the like, but even here the mental disturbance generally points to a peculiar temperament which predisposes to mental disease, and in a great many of these cases careful search will discover in near relatives insanity, or peculiarities approaching thereto. It is not every one who has suffered from sunstroke or has had a blow on the head that goes insane, and when we find upon inquiry that the majority of those who do, belong to neurotic families and have insane relatives, while the majority of those who do not are members of untainted families, there is but one conclusion to be drawn, viz., that even in what may be called traumatic insanity, hereditary predisposition cannot be ignored. Besides, it must be remembered that acquired diseased conditions of this kind tend to be transmitted to children afterwards begotten, as was the epilepsy artificially produced by Brown-Séquard in guinea-pigs transmitted to their young. Hence the man or woman who has been insane, be the cause what it may, can never be justified in becoming a parent. Even those in whose families general paralysis has occurred, should be most careful not to marry into a neurotic family, for this most hopeless disease, which was long looked upon as the result of fast living and dissipation in the healthy—indeed, in some of the most finely developed men—is, upon inquiry, turning out to be to a large extent confined to families where other mental and nervous diseases are common. In fifty-six cases of general paralysis in males, of which I have taken notes, I found a family history of insanity in no less than eleven, or 19.6 per cent., notwithstanding that in ten of the cases I could get no history whatever. The father had been insane in two cases, the mother in one, one or more sisters in four, a sister imbecile in one, and in three other cases near relatives in the direct line had been insane. Had I been able to inquire into all of the fifty-six cases, I have no doubt that the percentage showing hereditary taint would have been markedly higher.

In estimating the importance or gravity of the hereditary taint in any given case, several points are to be considered, as—whether the insanity has "run in the family" for some generations, or has appeared recently; whether the parent had been actually insane before he became a parent; whether one or both parents were tainted; the number of relatives in the direct line who have shown the neurotic temperament, and whether the disease attacks one sex only, as is sometimes the case, or appears equally frequently in both. From what has been said in the preceding pages, it should be clear, that the insanity of one parent would indicate a less degree of predisposition than that of a parent and a grandparent, for with each transmission the character gains prepotency; becomes more ingrained in the family nature, and so stands a better chance of being transmitted again. Insanity in uncles or aunts in the same line, added on to the disease in parent and grandparent, of course increases very greatly the gravity of the case; showing, as it does, the firm hold the disease has got upon the family.

A predisposition to insanity in both parents, even when it is not very deeply marked, is, as we have seen, always most serious. In such cases there is little or no chance of reversion to the healthy type. The children very frequently take on some of the more marked forms of nervous degeneracy, as idiocy or imbecility, with epilepsy or some physical deformity added on, as paralysis, club-foot, squint, blindness, deaf-mutism, &c. Even when neither imbecile nor deformed such children are very often sterile, and the family rapidly becomes extinct.

Here is the family tree of a patient of my own, which shows the terrible effect of a double parental taint upon the children :—

K. S.'s FAMILY.

M. Epileptic.	F. Had insane sister.

| M.
Epileptic.
Dead.
No issue. | F.
Epileptic and
insane.
Dead.
No issue. | M.
Idiot.
Impotent. | F.
Sane as
yet. | F.
Insane.
Suicidal melancholiac.
Incurable.
No issue. |

Has family of nine. Some are imbecile.

Here the epilepsy of the father, combining with the insane taint in the mother, came very nearly exterminating the family in one generation. Only one of the five wretched children leaves issue, and in all likelihood her miserable offspring are the last representatives of a decaying stock.

Many examples of the dire effect of the intermarriage of the neurotic or insane diathesis might be cited. I will, however, content myself with giving the following, which I take from Doutrebente's "*Annales Medico-Psychologiques,*" 1869.

First Generation. Father intelligent, became melancholic and died insane. Mother nervous and emotional.

Second Generation. Ten children. Three die in childhood. Seven reach maturity as follows :—Daughter A. a melancholiac ; daughter B. insane at twenty ; daughter C. imbecile ; daughter D. a suicide ; son E. imbecile ; son F. a melancholiac ; son G. a melancholiac.

Third Generation. A. has ten children ; five die in childhood, one is deformed, one has fits of insanity, one is eccentric and extravagant, two are intelligent and marry, but are childless. B. leaves no issue. C. has one child, a deformed imbecile. D. has three children ; one is an imbecile, one dies of apoplexy at twenty-three, and the third is an artist, described as "extravagant." E. has two children ; one dies insane, the other disappears, and is supposed to have committed suicide. F. is childless. G. has one child, who is imbecile.

In this family, as in that mentioned above, the presence of the neurotic taint in both parents rendered a reversion to the healthy type impossible, and Nature, refusing to continue a family so degenerate, stamped it out in the third generation.

Now, what would the world have lost that it could not well have spared, had the ancestors of these wretched families been forbidden the right of procreation? Nothing. It would have escaped an inestimable amount of suffering, past, present, and to come: a considerable amount of pauperism and consequent tax-gathering—that is all.

When the insane diathesis is present in only one parent, even though it be deeply marked, it is generally possible by wise marriages to lessen, and even perhaps in time completely to eradicate it. But when both parents are of the insane temperament the pathological character is so aggravated and deepened in the offspring, that they are never able to shake it off. Many of the children of such unions, as we have seen above, die in childhood. A great number of the remainder are sterile and deformed, and of those who come to maturity, and are fruitful, few indeed live in a second generation. The stock which springs from parents both of the insane type, almost invariably dies out in the second, or, at latest, third generation.

Again, the influence of the insanity of a parent in creating a predisposition in the offspring, will much depend upon the time at which the mental disorder has appeared, for while in every case its presence

shows a certain tendency to nervous disease or degeneration, yet if it did not appear in the parent
until after the offspring was begotten, its effect will
not be nearly so grave in the children as if the parent
had been actually insane before he became such.
Every attack of insanity, however brief its course,
increases the liability to subsequent attacks in the
individual, and also very greatly magnifies the danger
to the offspring afterwards born.

M. Baillarger, after careful research and the study
of a great number of cases, arrived at the following
conclusions, which have since been verified by several
observers and are accepted by most authorities.

" 1. The insanity of the mother, as regards transmission, is more serious than that of the father; not
only because the mother's disorder is more frequently
hereditary, but also because she transmits it to a
greater number of children.

" 2. The transmission of the mother's insanity is more
to be feared with respect to the girls than the boys;
that of the father, on the contrary, is more dangerous
as regards the boys than the girls.

" 3. The transmission of the mother's insanity is
scarcely more to be feared, as regards the boys, than
that of the father; the mother's insanity, on the contrary, is twice as dangerous to the daughters."

And now the question arises—which of these should
marry ? Who shall take the place of the censor and
say, this one and this shall, and these others shall not ?
In this very grave position the alienist physician often

finds himself, but his load of responsibility is generally lightened by the knowledge that his dictum is not final; that most of his clients have appealed to him in the hope that his verdict may be favourable, and with a determination already formed to act upon it if it be so, and to disregard it and risk the consequences if it prove the reverse. Too often he is told but half the truth—the applicant in too many cases is pleased to deceive himself and his adviser; and having so gained a favourable decision by fraud, deliberately enters on a course he knows to be studded with dangers; to live in a fool's paradise until the day of reckoning comes. In some cases it comes very soon, as where the first-born's vacant face is scanned day after day, and the heart sinks as the terrible fact forces itself upon the parent that his child is an idiot; or where the young wife suddenly loses all that made her godlike, all that made her human, and the husband finds himself with a creature in his arms at which his soul revolts. I have known a lady, young and beautiful, who within a month of her marriage was an inmate of a lunatic asylum, and who, though years have since passed, has not recovered, and in all probability will never return to home and husband.

But in many cases the evil day does not arrive until middle life; and then, when the fear once felt has worn away, when the deception practised has faded from the memory, and the grave admonition of the physician is forgotten, the son in whom the father hoped to live again, the girl on whom the mother's heart is set, is torn from the family circle a raving maniac, a tortured epileptic, a drunken criminal, or, happily, a suicide. Then arise sad regrets, but it is too late; the laws of Nature have been ignored, gratification has been purchased, and the price must be paid. The sins of the fathers shall be visited upon the children.

As to who should marry, it is clear that every case must be considered on its merits, and all cases in which there is even a suspicion of gravity should be submitted to the family physician, or some one specially learned in this class of disease. Happily it is not incumbent on the physician to advise celibacy in every case in which a member of a tainted family appeals to him, although unfortunately it is over the marriage of very few such he can pronounce the benediction of science. In the case of men there cannot be a doubt that in some cases the regularity and comfort of a happy married life would be of much benefit, and greatly reduce the liability to an outbreak of insanity, or to a relapse in one who had already been insane; but at best marriage partakes so much of the character of a lottery, that, unfortunately, it is impossible to say with any certainty whether it will in any given case prove happy or the reverse. Besides, it must be remembered that a man of insane temperament does not make the most patient and long-suffering of husbands, too often proving to be, like the sage, " gie ill to live wi'." However, if the taint be not too deeply marked, and more especially if the risk of progeny be not run, marriage with a suitable person should not be condemned. Of course in all such cases the second party to the contract should be made to understand the position of her future husband, and the risk he runs of at any time becoming insane.

But such permission can be extended to men alone, and to those only who have never been insane, and in whom the signs of the insane diathesis are not prominent. To those who have already had an attack of insanity, or in whom the insane diathesis is distinctly marked, it is impossible for the physician, having regard to the health and welfare of the community, to recommend marriage.

As to women, prohibition must be still more strict. The cares and trials of motherhood are so trying and severe that for her own sake no woman predisposed to insanity should be induced to add so greatly to her chances of losing her reason, while the terrible certainty with which the mother transmits her insane temperament to her children renders it impossible for the physician to consent to her putting herself in the way of becoming a parent.

This is all that can be done at present, namely, to give advice which possibly will be disregarded by the great majority of those to whom it is offered; but it is earnestly to be hoped that before long the Legislature will do something to stay in some degree the propagation of insanity. With our present knowledge of the hereditary character of this disease and its cruel and pauperising effect upon the populace, it is a scandal that persons who have been several times insane, and others who are at best but half-witted, should be allowed to marry and bring forth children to be a source of expense to the state and of contamination to future generations.

CHAPTER IX.

MARRIAGE AND DRUNKENNESS.

"The morbid craving for alcohol is common, and so intense that men who labour under it will gratify it without regard to their health, their wealth, their honour, their wives, their children, or their soul's salvation.

. . . After they have become dipsomaniacs, in the present state of the law that does not allow legal interference with their liberty —I say it with deliberation—the sooner they drink themselves to death the better. They are a curse to all who have to do with them. a nuisance and a danger to society, and propagators of a bad breed."—CLOUSTON.*

UNFORTUNATELY it is not necessary to say much in the way of proof of the transmissibility of the "drink crave," as this neurosis has been called. It speaks for itself from every grade of society in the land, from the highest to the lowest, and its voice gives forth no uncertain sound. With instances of the hereditary transmission of this curse every one is only too familiar, and I need not soil these pages with any long record of cases. All any one has to do is to look around among his friends and acquaintances to see how this sin in parents is visited upon the children.

The hereditary character of the abnormal condition

* "Mental Diseases," by T. S. Clouston, M.D.

of which habitual drunkenness is the outward sign, although firmly established and universally admitted, is not understood as it should be. It is too often looked upon as a vice acquired by the individual, the outcome of voluntary wickedness. In some cases this is doubtless true, but in the vast majority of cases inquiry into the family history will reveal the presence of an inherited taint, such families generally showing the neurotic or insane diathesis more or less distinctly marked. No grade in the social or intellectual world is, or ever has been, free from this disease, and if we study the family histories of the great ones of the earth who have fallen victims to it, we shall find that there the cause is the same as amongst the obscure, viz., that they have inherited a degenerate nerve-condition which renders them above others susceptible to this and allied neuroses, such as epilepsy, idiocy, madness, suicide, and the like. In fact, the dipsomaniac and habitual drunkard are very often as much sinned against as sinning, inasmuch as they have inherited an unstable nervous system which renders them liable at any time to fall victims to this vice under provocation which, upon a stable nervous organisation, would be powerless for evil.

Evidence of the hereditary character of this and other transmitted pathological conditions is seen in the tenacity with which they stick to their victims despite all treatment. An acquired vice or disease often gives way before persistent judicious treatment, but the innate evil is only to be eradicated by treatment carried on through several generations. Nevertheless, the physician's duty is to make the attempt in every case. In some few his efforts will be rewarded with more or less success, but, unhappily, in the vast majority they must end in utter failure, for the simple reason that he has been called in too late. As Dr. Oliver Wendell Holmes has said, "the doctor should have been called in a hundred years earlier."

By reference to Dr. Stewart's table, already given, it will be seen that he fixes the proportion of cases of dipsomania, in which he found hereditary taint, at so high a figure as 63.4 per cent., which is above that of any other form of mental disease given.

Of late years the Legislature has been induced to recognise habitual drunkenness as a diseased condition, and has made certain laws for the temporary confinement, care, and treatment of those so afflicted, if they themselves sanction it. This, however, is only a first step in the right direction, but the path is entered upon, and we may hope before long to be able to detain, as we now can a raving maniac, those unfortunate, semi-responsible creatures who at present outrage society by indulging their degraded appetites, and are free to propagate their innate degeneracy. And when that day arrives we shall enter upon an era in which it will be possible to lessen, in some part, not only habitual drunkenness, but all the diseases, mental and bodily, which arise from the abuse of alcohol.

I need not harrow the feelings of the reader by

submitting the terrible picture of the drunkard's home. It has been done in poetry and prose by many masters, and been depicted upon the canvas of the painter, and upon the stage, by some of our greatest artists. I shall not attempt to reproduce a picture so well known, but confine myself to the matter-of-fact statement that there is, perhaps, no disease or vice, hereditary or otherwise, which causes deeper degradation in the individual, more acute suffering in the family circle, or makes a greater call upon the ratepayer, than does this of habitual drunkenness. It is a curse upon the community, for it is the starting-point of insanities, epilepsies, crime, and endless disease in posterity, while as to the individual, there is no other diseased condition known which so utterly and rapidly destroys all moral sense; unless it be epilepsy, to which it is nearly allied. The victims of this horrible and irresistible craving may at first honestly express shame and regret for their weakness, and for the disgrace which they bring upon those who should be dearest to them. But this spirit is only too short-lived, soon the moral nature —never strong in such persons—becomes undermined, and we find the man or woman who but a short time before would have scorned dissimulation or untruth, transformed by his vice to a cunning, scheming liar, without the remotest sense of truth or honour, and ready to do absolutely anything to gain the where-withal with which to feed his thirst. Once on the down-grade, a man soon reaches a level where honour,

truth, and even common honesty are unknown, but
in woman the descent is even more rapid and terrible.
Once launched upon the downward journey her course
is not to be stayed. To every deep she finds a lower
depth; her home, her husband, her family, her very
honour, are, in turn, given a sacrifice to the demon
who is not to be appeased.

Yet after all that has been said and written on
this subject, these unfortunate creatures are still mis-
understood, and when their inborn vice leads them
beyond the lines laid down for the guidance of the
mass, they are haled before a court of justice, and
punished like the thief or other law-breaker. As
might be anticipated, this seldom if ever does any
good, and if proof of this were wanted, it would be
found in the regularity with which they return, time
after time, to their place before the judgment-seat.
Who has not come across such passages in the reports
of the proceedings at our police courts as these :—

" A shoe rivetter made his fiftieth appearance at the
police court this morning, when he was charged with
being drunk and disorderly ; and having, thanks to
the holiday-time, no money to pay for his 'jubilee,'
was sentenced to seven days."

"Margaret Hearn, who is better known as 'Mog
the fireman,' in consequence of her once having ascended
a fire-escape when under the influence of drink, and
excusing herself on the ground that she did so
'because she had heard of Jacob's ladder, and wanted
to get to heaven,' was charged, before Sir John Bridge,

at the Bow Street police court, yesterday, with being drunk and disorderly. 'Mog,' who entered the court in her characteristic style, said, 'Ah, Sir John Bridge, a happy New Year to you!' (Laughter.)—Constable 343 E stated that prisoner was found drunk in Drury Lane.—Bush, the assistant gaoler, said that the prisoner was a hard-working woman when sober, but could not keep from drink. She had several times been placed in a home, and on one occasion had stayed eleven months, and at the end of that period received a very good character. She had spent seventeen years out of the last twenty in prison.—Sir John Bridge thought that the best thing that could be done for the defendant was to put her in such a position that she could not have access to drink. He therefore ordered her to find two sureties to keep the peace for six months."—*Daily News*, 9th January 1891.

"Margaret West, a woman of the same class, who was said to have been before a magistrate upwards of fifty times, was charged with being drunk and disorderly.—Defendant (to Mr. Vaughan): 'It is all through the drink.'—Mr. Vaughan: 'Then why do you get so much to drink?'—Defendant: 'I am going to have a try this time.'—Assistant-gaoler Bush said the defendant had had many trials. She had only recently signed the pledge.—She was fined forty shillings, or in default twenty-one days' imprisonment."—*Daily News*, 27th January 1891.

Such persons as are here referred to are not responsible agents, and the state should recognise that

9

fact and act accordingly. These creatures are as helpless to fight against the desire for drink as is the hereditary suicide to fight against the fate which impels him to destruction, and their punishment is neither more just nor more beneficial than would be that of the epileptic for creating an obstruction by falling down upon the pavement. Justice will not be done until these " weak ones," instead of being packed off again and again to prison, and being permitted to propagate their kind in the intervals, are sent to some kind of industrial home or penitentiary where they will be guarded against temptation, where they may spend the full value of their labours in any comforts they please, except only intoxicants, and where the sexes shall be kept apart.

The fact that this drink crave is handed down through generations in most instances, can in no way justify any man or woman, however clean their family bill of health may be, in thinking that their indulgence in this vice will be harmless to their offspring. It must be remembered that acquired characters tend to be transmitted, and that the most vicious hereditary predisposition existent had a beginning in the healthy individual. Therefore, those who wish to live in posterity and see their children free from the mark of the beast—endowed with all the heaven-born attributes which raise man to his high position above all other creatures—must never, even temporarily, degrade their nature. True, one indulgence may not leave an impress sufficient to

appreciably affect the children. But that way danger
lies. An act once done, whether good or evil, is
easier to repeat for having been done before. The
appetite for alcohol is only too easily cultivated, and
the man or woman who, through weakness or thought-
lessness, saturates his brain with it frequently, must
not be surprised if his sin be visited upon his children
as idiocy, epilepsy, or other grave nervous or physical
deformity.

From the earliest times it has been known that
drunkenness is one of the most fruitful sources of
idiocy, and also of physical deformity and crime, in
the children. It will be remembered that it was the
drunkenness of Jupiter when Vulcan was conceived
to which was attributed the deformity of that god.
Dr. Howe, upon careful investigation, found that fifty
per cent. of all the idiots in the State of Massachusetts
examined by him were the children of intemperate
parents. Dr. Fletcher Beach sets down drunkenness,
either alone or associated with some other obliquity
of nature, as the cause of 25 per cent. of all the
idiocy received into the Darenth Asylum, and with
this estimate almost every other observer agrees.
When spoken of in this connection it is generally
chronic drunkenness that is meant, and certainly a
large part of the evil caused by the abuse of alcohol
arises from chronic or continued dissipation; nevertheless it must be clearly understood that a single
debauch may result in the idiocy or deformity of
the child then conceived. Cases are quite common

where a temporarily drunken person has begotten an idiot child. As Dr. Maudsley says, "Here, as elsewhere in Nature, like produces like, and the parent who makes himself a temporary lunatic or idiot by his degrading vice, propagates his kind in procreation, and entails on his children the curse of a most hopeless fate."

A striking illustration of the part played by drunkenness in the production of idiocy is to be found in Norway. In that country, in 1825, the spirit duty was removed, and, consequently, intemperance at once began to increase alarmingly among the people. The result—or rather one of the results—of this was, that during the first ten years following this regrettable event insanity increased among the Norwegians by 50 per cent. This was, perhaps, to be expected under the circumstances, but no one anticipated that the increase of congenital idiocy among the children born during the same decennial period would amount, as it did, to 150 per cent.

Drunkenness is one of the greatest—perhaps the greatest—agent of degeneration at work among the human race, and to it must be attributed much of the disease, crime, moral obliquity, and general degeneracy, physical, mental, and moral, which we find so common among the poorer classes in all large centres of civilisation. The dire effects of this agent of degeneration are to be found among almost every people upon the face of the earth, and in some countries they are simply appalling. In Sweden, for instance, which is one of the most drunken countries in the world, the people are deteriorating in a manner positively alarming. Some years ago Dr. Magnus Huss wrote of the Swedes:—"The whole people are degenerating: insanity, suicide, and crime are frightfully on the increase; new and aggravated diseases have invaded all classes of society; sterility and premature death of children are much more common; and congenital imbecility and idiocy are in fearful proportion to the numbers born."

Here is a history of a family which well shows the degenerating effect of drunkenness upon the stock.

First Generation. Father, a drunkard.

Second Generation. Son, a drunkard. Was disgustingly drunk on his marriage day.

Third Generation. Seven grandchildren. First died of convulsions. Second died of convulsions. Third was an idiot at twenty-two years of age. Fourth, melancholiac with suicidal tendency—became demented. Fifth, peculiar and irritable. Sixth has been insane repeatedly. Seventh nervous and depressed, and indulges in most despairing anticipations as to his life and reason.*

This drink-crave takes one of two forms, either habitual drunkenness, as seen in the toper, who is at all times when he can procure the drink more or less intoxicated; or dipsomania, in which the disease takes on a periodic character, breaking out at intervals of one to six or nine months, and rendering the individual

* "Traité des Dégénérescences" (*Morel*).

wholly irresponsible while the paroxysm lasts. These
two forms of the disease are totally distinct, the
paroxysmal seldom running into the habitual, or the
reverse. Indeed, they seem to attack persons of
altogether different temperaments, the toper being in
most instances a slow, obtuse, lethargic person both
in mind and body, with but little power of will, while
the true dipsomaniac is generally of quick excitable
nature, active and impulsive, and not infrequently,
before his disease has gone too far, of superior intellectual ability. In both cases, however, if life be prolonged the end is the same ; for while the dipsomaniac
is specially liable to sudden death from violence,
suicide, delirium tremens, &c., and the toper to disease
of such organs as the kidney, heart, liver, and brain,
yet if they be not so cut off each will arrive at the
same terminus, viz., gradual weakening of the mental
faculties terminating in complete dementia. In some
cases epilepsy, or some form of delusional insanity with
attacks of maniacal excitement, may precede the final
dementia, but dementia is the end.*

The distinction between these two forms of the disease is also marked in the progeny. The offspring of
the habitual drunkard generally inherits such degenerative conditions as idiocy, scrofula, deaf-mutism, the
tendency to phthisis, and sometimes epilepsy, while
that of the dipsomaniac is liable to the more active

* According to the returns of the Commissioners in Lunacy,
drunkenness is responsible for over 13 per cent. of all the insanity
coming under their notice during the ten years 1879-1888 inclusive,
20 per cent. of the males and 7 per cent. of the females.

or spasmodic forms of nervous disease, as suicide, acute
mania, epilepsy, and crime. The children of both are
peculiarly liable to convulsions, and death at an early
age from this cause is a frequent occurrence in such
families.

This diseased condition, like any other hereditary
predisposition, may remain latent for a generation and
reappear unexpectedly in the next, but it is seldom
that it does not show in some member of the family,
more especially in those children which were begotten
after the disease had been active in the parent; for,
as in other hereditary diseases, those children begotten
after the disease has declared itself by an acute attack
in the parent, are much more liable to inherit the predisposition than those born before such outbreak,
these latter appearing at times to escape the blight
altogether.

It is, perhaps, unnecessary to say that in this disease, as in the other neuroses, it is highly improper
that those in whom it is well marked should become
parents. They are unfitted by their inherited infirmity
to undertake the duties and responsibilities of married
life; as husbands or wives, and as parents, they are
equally sad failures. They are always improvident,
and their early death often saddles the community
with the care of a helpless family, while of the children it may be said that there is not sufficient chance
of their being useful to themselves or to the commonwealth to justify their being brought into existence.
Above all, there should be no intermarriage among

persons inheriting this disposition. If there be any person whose partner should be without taint it is assuredly him that carries within him the germ of such an insidious and degrading disease as this drink crave.

In this, as in the other insanities, the disease is much more dangerous in the mother than in the father, which is a sound reason why the daughters of drunken parents, often fascinating by their flighty, excitable, vivacious, neurotic manner, should be carefully avoided by men in search of mothers for their children. The man who marries the daughter of a drunkard, not only endangers his own self-respect and happiness, but entails to his children a wretched inheritance of degradation and suffering. On the other hand, no woman should be induced to marry a confirmed drunkard, and the disposition and character of the sons of such should be most carefully inquired into before any engagement is entered upon. This is one of the few instances in which a long engagement is not to be condemned, for frequently the engaged man loses that desire to appear well in the eyes of all women, which actuates most single men, and displays much of his real character.

Not a few of the best of our women throw themselves away, and ruin their whole lives, by marrying confirmed rakes and drunkards, in the hope, the almost insane hope, of saving them from the fate to which they have been foreordained by a bad inheritance. The spirit which prompts to such devotion and self-sacrifice is not to be treated lightly. An attribute so Christ-like is not to be rudely pushed aside by cold, calculating reason, without a word of sympathy. In some few cases, doubtless, men have been snatched as brands from the burning by noble women who have risked all in the hazard, and such wife-heroes should stand in the forefront of the ranks of Nature's nobility. Yet I would point out that the attempt so rarely ends in salvation, and so frequently in complete failure and despair, that such an experiment can in no case be advised; and further, that while one might not feel justified in interfering with attempts at the reclamation of the erring, if only the fate of the volunteer were at stake, he feels it his duty to speak when he remembers the children whose fate is also staked upon the hazard. It may be argued that a person has a right to risk happiness, even life itself, in the hope that some other may be benefited, but it cannot be said that a person should have legal or moral right to jeopardise the future of a whole family, to satisfy any instinct, however noble.

CHAPTER X.

MARRIAGE AND EPILEPSY.

"Epilepsy is pre-eminently an hereditary affection."—J. RUSSELL REYNOLDS, M.D.

EPILEPSY is one of the most fearful diseases which attack man. From the earliest times it has been more feared than even madness itself. Among the ancient peoples, the Jews, Greeks, and Romans, this disease was the foundation upon which was built the doctrine of demoniacal possession, and certainly the symptoms of the disorder, as observed during one of the terrible outbursts of maniacal fury to which epileptics are at all times liable, are enough to justify a belief in such a doctrine, in any ignorant and superstitious people. Dr. Clouston says, "No demon could by any possibility produce more fearful effects by entering into a man, than I have often seen result from epilepsy," and with this every one will agree who has witnessed one of those terrific outbursts, which so often convert such sufferers into very demons.

Although, in the majority of cases, not what might be called a rapidly fatal disease, epilepsy is one of the most cruelly painful and hopeless afflictions which can come upon man. It unfits the person it attacks, even from the earliest stages of its progress, for most of the duties of life. It is liable at any moment to render him unconscious, to precipitate him into fire or water, or from a height, while it may in a moment convert a rational being into a very angel of destruction. Like an evil genius it is ever present with its victim, lying in wait to strike him down at the most inopportune time and place. It may even, without the slightest warning, put an end to the life of its victim by the virulence of its attack, or, which is more common, by accident or suicide. This fact, however, is hardly to be deplored; indeed, such a termination is almost to be preferred to the fate which awaits the greater number of such sufferers, who in time sink into drivelling idiocy, or some of the most degraded and repulsive forms of vacuous dementia.

It may be truly said of the epileptic that he is doubly cursed. Racked in body, and robbed of mind, he soon becomes the most pitiable creature on earth. If he be so unfortunate as to survive, to escape all the pitfalls which his disease spreads in his path, too often he is only spared for a worse fate. All along he is liable to convulsions, which distort the features beyond recognition, convulse every muscle in the body, and rob him for the time being of consciousness. These attacks are usually followed by deep sleep, but they may be followed, preceded, or replaced, by outbursts of the wildest frenzy, which frequently takes a homicidal or suicidal form. When these outbursts occur, the epileptic will make ferocious and murderous assaults upon any one within his reach, and not a few slay relatives near and dear to them, or destroy their own lives, during such attacks. These outbursts may occur at any time or place. They may even appear before the intellect has been sufficiently weakened or disordered to attract attention, and for this reason no one who has ever been epileptic should be entrusted with the care of children, or put in any other position where opportunity is offered for the gratification of such homicidal impulse. The intellectual course of the epileptic between the fits—whether physical or mental—is always a downward one. Each attack leaves the whole mental apparatus more or less clouded and exhausted, and as this is never wholly recovered from, each attack marks a permanent impairment of mind—a step on the darkening path which leads down to intellectual death. In certain cases the maniacal attacks rarely appear, and the course of the disease is marked only by the convulsive seizures. But in these cases the result is the same. If life be sufficiently prolonged, the same dark destination is reached, whether the downward journey be a gradual descent, or be accomplished by a series of bounds, each marked by an outburst of murderous frenzy. In either case the goal is a degraded imbecility.

Epileptics are also specially liable to mental disorders of the more ordinary character, that is, mental disorder which continues in the intervals between the fits. They frequently, at an early period, suffer from auditory and visual hallucinations, and such cases are most liable to uncontrollable impulses, homicidal and suicidal, and to maniacal outbursts. Later, delusions more or less permanent are often developed, which are the cause of not a few of the motiveless murders constantly occurring. In fact, the majority of epileptics are really insane between the attacks. Esquirol * examined 385 epileptic women, with a view to discovering to what extent mental disease existed among them, and he discovered that only a sixth were free from intellectual derangement : " but nearly all these," he said, " were irritable, peculiar, and easily enraged." From this it is clear that no person who has ever had " fits " is to be depended upon, and to entrust young children or others to their care, as is often done by mothers and others in authority out of misplaced pity, is little short of criminal.

It is most unfortunate, then, that this dread disease, which degrades its victims to a level beneath that of the beasts that perish, is not properly understanded of the people. Not only the ignorant, but a great number of those who can lay claim to some education, know so little of hereditary disease that they look upon epilepsy occurring in children merely as an un-avoidable affliction depending entirely upon teething, worms, or some such infantile trouble, which for some unknown reason is often continued into adult life. In the same way, when it makes its appearance later in life, it is set down as the result of a fright, the " change of life " (puberty), or some such cause. Not

* " Maladies Mentales," vol. i.

once in a hundred cases is it attributed, as it should be, to inherited predisposition. This, I say, is unfortunate, for ignorance of the true character of the disease can have but one effect—its propagation.

The people should be taught that epilepsy is, *par excellence*, an hereditary affliction, that it is nearly related to idiocy, and madness, and paralysis, and deaf-mutism, and that no member of any family in which it is known to be should be considered a person who can with safety become a parent. Of course, there are cases of epilepsy which arise from mechanical injury to the head, or fright, or have their exciting cause or starting-point in such irritation of the nervous system as that experienced during dentition or the approach of puberty. Yet it must never be forgotten, that even in these cases much depends on family taint, and that epilepsy seldom attacks a child who is not the unfortunate inheritor of a neurotic temperament. Indeed, were it otherwise, most children must be epileptic, for all suffer the nervous irritation and excitement consequent on dentition and the approach of puberty and adolescence, and if all were equally predisposed, all must equally suffer. But while all suffer these irritations few become epileptic, and it is perfectly clear from inquiry into the family histories of those who do, that they are predisposed by inheritance, else they too would escape.

Epilepsy is, in fact, one of the most strongly hereditary of all diseases. In this respect, it is on a footing with the suicidal impulse, melancholia, drunkenness, and gout. Dr. J. Russell Reynolds found heredity well marked in 31 per cent. of his cases, and says, " I am therefore led to believe that an hereditary tendency to epilepsy is much more common than it is generally represented to be by recent writers on the subject." Echeverria said 28 per cent. of all the cases coming under his notice were hereditary. Webster in England, and Esquirol in France, declared that a third of all cases of epilepsy depended on family taint, while Dr. Gowers, one of the greatest authorities on the subject, asserts that no less than 36 per cent. of all epilepsy has hereditary transmitted predisposition as a foundation.

I myself have records of 143 consecutive cases of epilepsy, as they appeared for admission into an asylum for the insane. There were 93 males and 50 females. Of the males, 34.4 per cent. were members of families in which either epilepsy or insanity of some description had already appeared; of the females, 50 per cent. belonged to the same class; while in 39.8 of the total of both sexes there was positive evidence of hereditary taint. I would also remark that in a considerable number of my cases, no history of any kind could be obtained.

Although epilepsy is frequently transmuted to insanity, idiocy, chorea, hysteria, the drink-crave, and other diseased conditions of the nervous system in transmission, it is remarkable for the regularity with which it is transmitted unchanged in some families. Dr. Gowers found in nearly 75 per cent. of all his

hereditary cases, either epilepsy alone, or epilepsy
combined with insanity of some kind in the direct
line of ancestors, and instances a case in which no
fewer than fourteen members of a family were afflicted
with the disease. Yet notwithstanding the fact that
epilepsy is thus transmitted unchanged with wonderful regularity—almost approaching in this respect the
suicidal impulse—the taint not infrequently takes
other form in the offspring. As Dr. Clouston remarks,
" Hereditarily ordinary insanity and epilepsy are closely
allied. The son or daughter of an epileptic is just
as likely to be idiotic, weak-minded, drunken, or insane, as to be epileptic; and certainly, the children of
families with strong insane heredity are very commonly epileptic." * In truth, there are few families
showing the insane diathesis which have not epileptics
among their members. When the tendency to insanity becomes deeply marked in a family, epilepsy
with idiocy, paralysis, squint, physical deformity, or
deaf-mutism added on, is almost certain to make its
appearance in some of the children, and when this
stage of degeneration is reached the extinction of the
family is at hand, such children being generally sterile.

Two families of which I have notes are good
examples of the fatal effect of insanity, combined
with epilepsy, upon the family. One is that already
mentioned at page 97, where, out of a family of nine
children, six died of "fits" during the first year of
life, and of the survivors, one was an idiot and the

* *Loc. cit.*

other two were "weak"—*i.e.*, imbecile. Of that family there will never be another generation. The second was the family of an epileptic, who married a woman of the insane diathesis. The family tree is given at page 108, and shows the disastrous effect of this parental combination upon the offspring.

In this case the epilepsy of the father, combined with the insane taint in the mother, came very nearly reaching the necessarily fatal type and exterminating the family in one generation. Of the five wretched children born, the first was epileptic, the second epileptic and insane, the third an impotent idiot, and the fifth insane—a melancholiac with strong suicidal tendencies. In fact, the family was so saturated with nervous disease, that the end was almost reached.

It must not be imagined that the above family records are in any way exceptional; unfortunately, they are only too common. I take the two below from Dr. Fletcher Beach, of Darenth Asylum :—

CASE I.

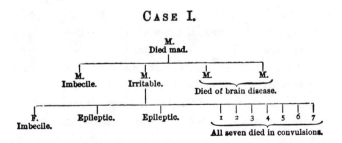

Here the father, in whom the family taint was represented by mere irritability, begot a family of ten who

10

were probably the last of a degenerate race. Of the ten, seven died in childhood of convulsions, two were epileptic, and the remaining one was an impotent imbecile.

CASE II.*

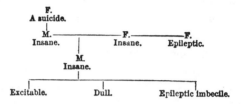

Here, again, we are within measurable distance of, if not actually arrived at, the close of the family's existence. Had the man, the son of the suicide, had the good sense or good fortune to have married a woman of sound family, instead of the sister of an epileptic, who was to become insane herself later in life, it is possible that reversion to the healthy type might have taken place, at least in some of the children, and the fate of the family been other than it was.

From all these family histories it is evident that epilepsy is symptomatic of a lower grade of degeneration than that found in the ordinary insanities. It most frequently makes its appearance in members of families which have shown, or are showing, other phases of degeneration, as idiocy, drunkenness, defor-

---

* *British Medical Journal,* May 28, 1887.

mity, criminality, cancer, scrofula, and insanity. Few, indeed, are the families which become extinct without displaying epilepsy in some of their latest representatives. It is the constant associate of idiocy, which is the lowest form of development consistent with a continuance of life. M. Herpin, the distinguished French physician, calculated that epilepsy occurred in six out of every 1000 of the general population (and even this was thought by some to be too high an average); whereas Drs. Ireland, Langdon Down, and others found that close on 25 per cent. of all idiots are epileptic, that is, 250 to the 1000 against six to the 1000 among the ordinary population. The frequency with which epilepsy attacks the drunken, the instinctive criminal class, the scrofulous, and the insane is notorious. From the figures of the Commissioners in Lunacy we find that 90 to the 1000 is the rate at which it is to be found among the insane of these countries, while the studies of Lombroso and others have proved pretty clearly that anthropologically the epileptic and the criminal are very nearly akin.

In the great mass of cases epilepsy makes its appearance during childhood, this being the more certainly true where the disease is distinctly hereditary. Dr. Russell Reynolds found that 83.3 per cent. of hereditary epilepsy appeared under fifteen years of age, while of the non-hereditary only 46.1 per cent. appeared under that age. He also expressed the opinion that the disease appears earlier in life the more strong and direct the taint. He gives the average age at which it appears in the male as fourteen, and in the female eleven years.

According to the Registrar-General's returns the average number of deaths annually registered as due to epilepsy during the past ten years has been considerably over 2500. These figures, however, can give no idea of the prevalence of the disease among the population, for it is a very small proportion of those who have been epileptic whose deaths are directly due to that disease, and fewer still which are registered as being so. Not a few epileptics come to their death by violence, either accidental or suicidal, to which they are more liable than any other class of the community, and it is well known that a vast number of epileptics annually succumb to such degenerative diseases as phthisis. Besides, how many of the enormous number (20,000 to 25,000) of deaths annually registered under the head "Convulsions" are due to inherited epilepsy it is impossible to say; that a very great number are is absolutely certain. Perhaps if we were to assign to inherited epileptic taint a half of these so-called deaths from convulsions, and say that about 12,000 deaths are annually due to this disease, we should be well within the mark. And of these 12,000, the majority would be those of tender years, whom inheritance of a neurotic temperament had doomed to an early death. The amount of infantile suffering represented by these figures can only be surmised.

And now arises the question of marriage. One thing is certain, and that is, "epileptics decidedly ought not to marry." Great numbers of these afflicted ones are at present detained in asylums and workhouses, but this at present only can be done when the disease has gone so far as to affect the mind sufficiently seriously to render the sufferer insane in the eye of the law, and before this stage is reached the epileptic is frequently the parent of a family.

Laws aiming at preventing epileptics becoming parents have been known in times past,* and seeing to what an extent our idiots, criminals, suicides, drunkards, and insane are recruited from the offspring of the epileptic, I think the Legislature would be fully justified in forbidding the confirmed epileptic becoming a parent, as a proceeding inimical to the weal of the commonwealth.

Legislation to this end is, perhaps, too much to expect just at present, but the day when such a law will appear on the statute-book is fast approaching. The divine right of kings to govern, once as firmly fixed as any canon of the Church, has disappeared before the onward march of education and enlightenment, and so shall what some are pleased to call "the divine right of procreation." It may be said that it is not necessary to interfere; that if we leave the whole affair to Nature she will right herself. Undoubtedly. But that is exactly what we do not do. If left to themselves, these wastrels from Nature's workshop could not survive, they would succumb to their own unfitness. But this happy consummation

* Boethius, " De Veterum Scotorum Moribus," lib. i.

we use every endeavour to postpone. Nature will not
even tolerate the unfit: we not only cherish them,
but strive to cultivate them, and the result of our
labours is to be found in our asylums for the blind,
the deaf, the idiotic, the insane, and in our prisons.
By all means let us comfort and protect the helpless,
it is our duty: but let us stop there. If the course
of Nature were not interfered with, legislation would
be uncalled for.

For the present, however, we must do the best we
can to stay the ravages of this avoidable suffering
and poverty by education, by pointing out to the
people that the man or woman who knowingly takes
for husband or wife one so diseased as to rob the
children of a reasonable chance of health, is com-
mitting an outrage against God and Nature.

CHAPTER XI.

SYPHILIS cannot rightly be called an hereditary disease, for the reason that it is not handed down from parent to child through many generations, nor does it ever skip a generation to appear in that following. Indeed, it is still doubtful whether syphilis *as syphilis* ever reaches the third generation. But although the disease is seldom or never transmitted as syphilis beyond the children of the infected parent, it is often the starting-point of degenerate conditions which are transmitted through many generations, and cause grave deterioration in the family stock. On this ground, then, it deserves brief consideration in these pages.

Syphilis is a disease which affects the whole system. No tissue or organ is safe from its attack. When severe, and still more surely when engrafted upon a scrofulous, neurotic, or otherwise already degenerate constitution, this disease so impoverishes the system, robs the tissues and organs of their vitality, that a condition allied to the scrofulous diathesis is established. This condition is transmitted to the offspring

—often without a single symptom of the specific disease—and there forms the basis of epilepsies, scrofula, idiocy, physical deformities, and other pathological characters in them and their offspring. Until recently it was believed by many that the scrofulous diathesis in the child was but another form of the syphilitic cachexia of the parent. The recent discovery of the tubercle bacillus with its distinctive characteristics, proves conclusively that there is no kinship between syphilis and tubercular disease, but it in no way affects the belief that the impoverished condition of the syphilitic and his offspring is almost identical with that condition known as the scrofulous diathesis. Certain it is that the miserable devitalised children of the syphilitic offer a peculiarly favourable field for the growth and development of disease germs, and as that of tubercle is one having a special affinity for the tissues of those low in vital energy, a great number of such children fall victims to tubercular disease. As a factor in the production of general deterioration of the family, no other agent at present known can be compared with this disease—syphilis.

From the time of the formation of the primary sore at the seat of inoculation of the syphilitic poison, or perhaps earlier, up till two or three years after the last signs of what are known as the "secondary symptoms," the person infected will transmit the *disease itself* to any child born or begotten, and in cases which have been neglected—in which a strict and scientific course of treatment has not been carried out—the infected person may, even after ten, twelve, or more years, have syphilitic children. While in some instances the disease poison takes such hold upon the system as to rob the individual altogether of his or her highest function, viz., that of continuing the species.

There is no other disease the poison of which has so malignant and lasting an effect upon the constitution of man as syphilis, and the person who has been once infected, is never safe either as an individual, or as a parent. He may, certainly, after a course of careful treatment enjoy good health and beget healthy children, and imagine himself clear of the disease and its evil effects; but even after years he may beget a tainted child, or, if he catch some other disorder, its course may be affected, and grave complications arise because of the old taint. Excesses on his part will be punished by prolonged attacks of ill-health, and injuries which would speedily heal in the healthy, in him will give rise to tedious and exhausting ulcerations, diseases of the bones, and a hundred other distressing troubles.

We may briefly consider this disease under two heads, as Acquired Syphilis and Hereditary Syphilis, at the same time glancing at the degenerate conditions we have hinted at as following in its wake.

ACQUIRED SYPHILIS.—In order to acquire this disease it is necessary that a quantity of the living poison be introduced into the system. The quantity may be infinitesimal, and is generally introduced through some cut, scratch, abrasion, or other break in the cutaneous or mucous surface. Of course in the vast majority of cases the poison is received during unclean sexual connection, but the public should understand that the disease may be, and at times is, acquired in other ways, as we shall see later.

About three to four weeks after the introduction of the poison, a sore, having decided characteristics of its own, appears at the seat of infection, and soon after the lymphatic glands in the neighbourhood of the sore become enlarged. These do not suppurate, nor are they painful. Later the glands in distant parts of the body may also become enlarged, and this is a distinctive sign that the disease is true syphilis.

The primary sore generally heals without much trouble, and about six weeks later what are called the " secondary symptoms" appear. The person feels out of sorts, becomes feverish and has pains in the head, and shortly a rash appears on the skin in the form of red spots, or papules. The rash may be very plentiful all over the body, or may consist of only a few spots, but its plentifulness or the reverse cannot be taken as any criterion of the severity or mildness of the attack. About this time the throat becomes painful and inflamed, and sores form there, the hair gets thin—in fact, may all be lost, and intractable affections of the nails may appear.

At this stage the patient is most dangerous to those around him. The discharge from the sore throat, or from any other sore upon the body, even the blood itself, being loaded with the poison, which has multiplied a thousandfold in the system since its introduction, will convey the disease. It is therefore obvious that the very greatest care should be taken to prevent the spread of the poison. All cups, spoons, and other utensils used by such persons should be thoroughly cleansed before being used by others; all dressings, rags, &c., contaminated by contact with the sick one, should be burnt, and intimate relations with him generally should be strictly avoided. Kissing is a common means of conveying infection, and the use of the same cup, spoon, towel, &c., is not infrequently attended with a like result, consequently the greatest care is necessary in every case, and as from the nature of the disease the patient's friends and relatives are rarely aware of their danger, the duty of observing every possible precaution devolves upon the infected one himself.

The rash upon the skin, the sore throat, the dropping of the hair, together with various other troubles, make up what are known as the " secondary symptoms," and until the disappearance of the last of these the danger of infection to those around continues. Any child begotten or borne during this period, or within from two to three years after the disappearance of these symptoms, is almost certain to be diseased, however carefully the affection has been treated.

After the disappearance of the secondary symptoms, if the disease have been carefully and methodically treated from the first under the advice of a competent medical man, comparatively good health may be once more attained, and with attention to the laws of health this desirable condition may be continued, perhaps, through life. But if the person who has suffered from syphilis violate the laws of health, even in a mild degree, his apparent good health will almost assuredly break down, and he will become the victim of most intractable and distressing disorders, both internal and external. These are called the " tertiary symptoms," and often depend upon a diseased condition of the internal organs, in which the slowly working disease poison has caused the growth of masses of lowly organised tissue which interfere with the proper functions of the organs affected.

Other symptoms, less grave, perhaps, but no less distressing, may also make their appearance about this time, such as ulcerations of the cutaneous and mucous surfaces and diseases of the bones. It is therefore of the greatest importance that the person who has had syphilis should give strict attention to all hygienic laws, and make every endeavour to keep his health at as high a level as possible.

All persons, however, are not equally liable to the later symptoms of syphilis. " The constitution of the person will materially influence the phenomena which supervene during syphilis, *e.g.*, the gouty, rheumatic, tuberculous, and cancerous temperaments will modify the syphilitic lesions and degenerations; while constitutional syphilis in its turn modifies the character of ordinary diseases."* Such degenerate constitutional

* Sir William Aitken, *op. cit.*

states as those mentioned render the system less able to withstand the onset of so malignant and impoverishing a disease as syphilis, consequently it runs a more terrible and destructive course; the new poison naturally attacks the weakest tissues in such enfeebled constitutions, or rather, it seems to combine with the already existing constitutional disease to evolve a deeper degeneration. Thus in the gouty the poison attacks the joints and the great blood-vessels, and "the lesions ultimately assume the form of degenerations;" in the rheumatic the tendons, joints, eyes, and bones suffer most severely; in the scrofulous those tissues and organs in which ulceration is most liable to set up, or where tubercles are most likely to deposit, are most deeply affected, and so on with other family degenerations.

On the other hand, the enfeebled, impoverished, and diseased condition which is invariably produced by syphilis, renders the system infinitely more liable to the attack of other diseases. And when these do occur their course is perverted, grave complications arise, and the chances of some chronic ailment being left behind, or of the illness proving fatal, are largely increased.

HEREDITARY SYPHILIS.—As we have seen, true syphilis may be transmitted from parent to child. In every case this may take place for a period extending over two to three years from the time of infection, and in some cases even after the lapse of ten, fifteen, or more years. True hereditary syphilis is that which is transmitted from one or other parent *at the time of conception.* A child may be infected with syphilis at any period between conception and birth, by a mother who has during that time acquired the disease, but in such cases, although the child is syphilised before it is born, the disease is not true hereditary syphilis, nor is it nearly so deadly as where it is, so to speak, begotten with the germ.

Hereditary syphilis is one of the most common causes of abortion, and is also the cause of an enormous number of the deaths which occur during the first year or two of life. The virulence of the poison is often so great that the child is killed shortly after conception, and is born dead. In other cases, where the poison is a shade less virulent, the child is born with the disease actually upon it, and often only comes into the world to be carried to the grave. Others, again, are born without active symptoms of the disease, wretched, puny creatures, who develop the disease within a short time of birth, and of whom a large percentage succumb after a short and wretched existence. Another class are those in whom the poison seems to have left a general blight; the vitality of these is at so low a level that their development— physical and mental—is not only retarded but limited. Such creatures develop slowly; at eighteen or twenty years of age they are still children, and such they remain during life. Still another class show the evil influence of their parents' disease in their own deformity. These have malformed limbs, club-feet, hare-lip, cleft palate, spina-bifida, water on the brain, paralysis, and various forms of idiocy. Some may think that in charging these deformities to any large extent against parental syphilis, I am going too far. To these I would point out that it is not an assertion founded upon my own comparatively limited experience, but is the view held by Professor Fournier, one of the greatest authorities upon syphilis which the world at present possesses, and of many others only less eminent.

In syphilis the nature of the disease forbids our ever getting a full and honest history of its effects upon the offspring, and for this reason statistics on the subject are rare. Of its terrible effect upon the children we have, nevertheless, ample, if fragmentary evidence. The best that I can at present lay my hand upon is that published by Dr. B. Tarnowsky (*Der Kinderarzt*, October 1890). This distinguished observer takes a most gloomy view of the effects of syphilis upon succeeding generations. According to his experience, 71 per cent. of women suffering from syphilis either give birth to dead children, or bear children which die within a year of birth. This high percentage closely agrees with that of Professor Fournier, which I shall give presently. In his interesting and able paper, Dr. Tarnowsky records the terrible history of three families, whose fathers had contracted syphilis six, five, and four years respectively before marriage. All these men appeared to be cured when married, and all their children were born healthy, that is, they showed no symptoms of syphilis. In these three families there were twenty-two children, and of these only one grew up to healthy maturity. Five were premature, three died of inflammation of the membranes of the brain before attaining their second year, two were imbecile, two were idiotic, one had numerous signs of degeneration, one was weak in intellect, one insane, two hysterical, one epileptic, one a deaf-mute, and two had water on the brain. Of the thirteen still alive when these statistics were taken, eight were incapable of earning their living, the remaining five being sickly and nervous. All three families, he points out, were of the respectable commercial class; none of the children were exposed to the hardships which, in the case of peasants and artisans, may cause infantile diseases falsely attributed to syphilis. Dr. Tarnowsky has collected other family histories scarcely less dreadful, and his conclusion is, that syphilis in a parent may be the cause of a long series of most serious diseases — scrofula, rickets, nervous disorders, &c., and the offspring at the best are often weak, useless members of society.

All syphilitic children are ill-developed, miserable, puny things even when not deformed. Their little faces are withered, pale, and pinched; their noses become flat, their heads are large, their foreheads square, their cheeks seared with the scars of old sores; and over all there is a strange uncanny look of age and suffering which is repulsive, and strangely at variance with the cherub-like freshness and innocence of the healthy infant.

These wretched children, being robbed of a large part of their vitality, are predisposed to all sorts of disease. They are peculiarly liable to diseases of the nervous system, and a great number of them die from inflammation of the membranes of the brain and from convulsions, while the number that succumbs to tubercular disease—more especially of the bones and joints, as seen in spinal and hip-joint disease—shows how good a field their weak nature offers to all the micro-organisms of degeneration and death.

But although death is so busy in the ranks of the hereditarily syphilitic during the early years of life, many of them drag through a more or less miserable existence to what is to them maturity. Among these the standard of development, physical and mental, is generally low. In middle life—they never reach old age—they succumb rapidly to acute febrile disease, or the nervous system gives way and they become insane or epileptic. Under circumstances which would prove harmless to the robust, they develop phthisis and various scrofulous disorders, and to their offspring they transmit their degenerate natures and so deteriorate the race.

In some cases the child who has inherited syphilis does not show any outward or active sign of it in the early years of life, and occasionally it is not until the fifth, tenth, fifteenth, or even twentieth year that the inherited disease wakes up to activity (Fournier). Such cases are, however, rare. In the vast majority of cases where syphilis has been inherited, the dis-

11

ease makes its appearance during the first year or two of life, and either kills its victim, or leaves him feeble and blighted, an easy prey to some other affection.

As might be expected, all the graver forms of degeneration following or arising from syphilis are to be found at their best—or rather worst—among the poor in our large towns, where the disease is neglected, and where it has so many other agents of degeneration at work to help it. The ravages of syphilis among the poor of our great cities has long been recognised, but its influence as a degenerating agent among the lower classes has not yet been fully appreciated. Aided by drunkenness, poverty, and squalor, syphilis is largely responsible for that residuum of humanity to be found in the dark places of our great centres of population, from which are recruited the consumptive, the scrofulous, the epileptic, the prostitute, the idiot, the habitual drunkard, the instinctive criminal, and the insane.

The fearful malignancy of the syphilitic poison is best seen among semi-barbarous tribes where it has been introduced, and where no treatment mitigates its virulence. In such cases, a healthy people will soon become a degenerate, disease-ridden race, their constitutional condition often becoming nearly allied to that of the leper. Indeed, some of our very highest authorities have recently gravely questioned whether leprosy be not really one of the disease degenerations which follow neglected syphilis. At present, syphilis, with its following of degenerative disease, bids fair to exterminate the inhabitants of some of the South Sea Islands.

As I have said, hereditary syphilis may be transmitted from either parent. Statistics go to prove that in a majority of cases the infection comes, as we might anticipate, from the father. As to whether that of paternal or maternal origin be the more virulent, opinions differ, but at present the weight of evidence appears to be strongly in favour of the maternal being the more fatal. Professor Fournier found the inherited disease to be *fatal* to the offspring in the following varying proportions according to its source:—Paternal 28 per cent., maternal 60 per cent., and where both parents were syphilitic, the mortality reached 68.5 per cent. of the pregnancies.

Not only, then, is this disease the cause of an enormous amount of ill-health, suffering, and family decay among every rank of our population, but it is also the cause of death in thousands of children, born and unborn, annually. " In London alone, during twelve years (1854–65), there were 3370 deaths from syphilis among children, and 2587 of these were under a year old." How many children were killed before birth by this disease during these years can only be surmised, but that the above figures represent only a very small proportion of the total of the child life so destroyed we may be absolutely certain.

As to the source of this disease it is unnecessary to say a word. It is known to all that prostitution is the poisoned fountain from which this contamination is continually being carried to the people. Prostitution is older than history, and has existed in all nations and among all peoples; only varying in degree as the laws affecting the relationship of the sexes varied. All attempts at the suppression of this vice have signally failed; and it is to be feared that so long as human nature remains what it is, future efforts in that direction are foredoomed to like failure. Prostitution is but the vicious overflow of that passion which is the foundation of family and home, and which calls forth all that is best and highest in our nature, as well as much that is bad. So long as that passion exists, so long will this evil continue. That fact we must recognise, and having recognised, the duty of society and the state is clear. It is not to close our eyes against the evil we find it impossible to remedy, and, ostrich-like, try to believe that what we do not see does not exist; but, having admitted the incurable nature of the evil, to set about mitigating its evil effects in every way that is possible. How this social vice might be shorn of its terrible effect upon the guilty and the innocent alike, has been pointed out again and again, but until hypocritical self-righteousness takes a less prominent place in our creed, science will not be permitted to limit the ravages of this disease in England.

"In the eyes of every physician, and, indeed, in the eyes of most Continental writers who have adverted to the subject, no other feature of English life appears so infamous, as the fact that an epidemic, which is one of the most dreadful now existing among mankind, which communicates itself from the guilty husband to the innocent wife, and even transmits its taint to the offspring, and which the experience of other nations conclusively proves may be vastly diminished, should be suffered to rage unchecked, because the Legislature refuses to take official cognisance of its existence, or proper sanitary measures for its repression."* Infamous indeed! The historian has hit upon the proper word.

As to preventive measures, the one great law is— Be chaste. Lead a pure and virtuous life, and fear not this evil. As to accidental infection, the greatest care should be taken when it is known or suspected that any one in the household is suffering from the disease. The use of the same cups, spoons, towels, tobacco-pipes, &c., &c., should be carefully avoided, and the act of kissing should never be indulged in. In the case of a syphilitic infant, the parents should not permit friends to kiss it—should any be so inclined—and they should on no account engage for such a child a wet-nurse. Virtuous women, the wives of respectable men, often become infected from diseased children they have been engaged to suckle. No medical man will sanction such a proceeding as the wet-nursing of a syphilitic child. To submit any person to such grave danger of infection is a heartless and cruel outrage, and when wittingly committed should be a criminal offence.

* Lecky's " History of European Morals."

On the other hand, too great care cannot be taken in seeing that any wet-nurse engaged for a healthy child is not infected with this disease, for, as the diseased infant can convey it to the nurse, so can a diseased nurse convey it to the healthy infant. Parents should never engage any person as wet-nurse until their own medical attendant has examined and approved of her.

As to the danger of infection through vaccination, which has been vastly magnified by the opponents of that operation, little need be said here. That syphilis can be, and has been conveyed by vaccination, is unfortunately true; but as the operation of vaccination is always in the hands of qualified medical men, I have only to say that, with even reasonable knowledge and care, such an accident is next to impossible, and that the vaccinator who conveys syphilis should in all cases be held responsible.

In the next place, I would point out that there is no disease known in the course of which more absolutely marvellous results can be attained by good treatment, or more disastrous results by bad, than in syphilis. For this reason, the person who has any reason to suspect that he has been infected, should at once apply to a respectable medical practitioner, avoiding as he would poison all those ignorant vultures who advertise their nostrums, and feed and grow fat on the weak, the credulous, and the ignorant. A considerable number of the very worst cases which turn up in the hospital and the consulting-room, are those which have been altogether neglected, or, what is even worse, maltreated by advertising charlatans. Shame, and fear of exposure, drive many of the erring into the net spread for them by these quacks, and once there, escape is more difficult than might be expected ; in fact, to any but the strong-minded—few of whom are to be found in the position—it is next to impossible. Many of these ruffians, when they have learnt the name, position, and resources of their victim, proceed in the most deliberate manner, by the aid of threats of exposure, to levy black-mail. Thus the person who puts himself into the power of such sharpers is liable to financial as well as physical ruin ; the secret which would have been absolutely safe in the keeping of any medical man being made the lever for extorting sums which often exceed what would suffice to fee the most eminent specialist that could be chosen.

In cases of inherited syphilis, the unfortunate child should at once be placed under medical treatment. Such children will require, during infancy and youth, and, indeed, all through life, particularly jealous care and treatment to secure a reasonable development, and prevent, if possible, the onset of epilepsy, scrofula, phthisis, or some other disease of degeneration.

As to the question of marriage, experience has shown that it is not safe for the person who has been syphilitic to marry for a considerable time after the reception of the poison. Mr. Jonathan Hutchinson, our greatest authority in England on the subject, lays it down that marriage may be permitted " two years after the disappearance of the last of the secondary symptoms." This would be about two-and-a-half years after the time of infection. Professor Fournier extends the time of prohibition to " three to four years " from the time of infection, and certainly safety is on that side. Of course, these rules only apply to those cases which have been carefully and skilfully treated.

Finally, no person who has had syphilis should ever, under any circumstances, marry without the sanction of his medical adviser.

CHAPTER XII.

DEAF-MUTISM.

"Ordinary deaf-mutism is closely allied to idiocy, and is one of the hereditary neuroses. To me it is a physiological sin that marriages between such persons should be legal."—CLOUSTON.*

THERE is still some difference of opinion as to the hereditary character of deaf-mutism, and some might even demur to its being given a place in a work treating solely of hereditarily transmitted affections. This difference of opinion depends largely upon the want of discrimination with which cases are collected and classified. To understand the part played by family predisposition in the production of this condition, all cases of deaf-mutism should be divided with great care into two classes. First, those in whom the deafness is the result of congenital defect; and second, those in whom deafness follows some injury to the auditory apparatus before or shortly after the power of speech has been attained. Nearly the whole of the cases making up the first class we shall find dependent upon parental taint; whereas in the second

* *Op. cit.*, p. 286.

class the deafness, and consequent ignorance of spoken language (except in those cases in which the deafness is caused by scrofulous disease, syphilis, &c.), are no more hereditary than is the blindness of the child whose eye has been poked out with a stick. The two classes should be kept distinct. The first is by far the larger, but a sufficient number belong to the second to materially affect statistics; and until the two classes are clearly distinguished, opinions will differ as to the part played by hereditary taint in the production of deaf-mutism.

With the second class we have nothing here to do. Such cases generally arise from some destructive inflammatory affection of the middle ear—as that which frequently follows scarlet fever—and so may occur in any child. In the infant they are easily distinguishable from the congenital cases; but later in life, when the history becomes garbled, it is often difficult, if not impossible, to distinguish to which class a case properly belongs.

Congenital deaf-mutism is, on the other hand, a constitutional affection. It is a degenerate, a markedly degenerate condition, and therefore a sign that the family in which it appears has started on the down grade toward decay, except in such cases as it depends upon some temporary depraved condition in the parents at the time the child was begotten, as drunkenness, &c.

It is true that deaf-mutism is not, like the suicidal impulse and some other conditions that we have con-

sidered, transmitted in a majority of cases unchanged
from parent to child. The family defect which shows
in the child as congenital deafness may be met with
in the ancestors or collateral relatives as idiocy,
insanity, blindness, epilepsy, scrofula, physical deformity, or the like, for these are but the various outward signs of that general tendency to degeneration
which mark such families. It is not necessary to
trace deaf-mutism as deaf-mutism through several generations of the family line to prove its dependence on
hereditary taint. It is but the sign by which the
innate degeneracy makes itself known to the outer
world, and this sign will vary in successive generations,
and even in different members of the same generation.
Thus in the family of the deaf-mute, inquiry will
frequently discover idiotic, epileptic, blind, or scrofulous
brothers and sisters; dipsomania, insanity, epilepsy,
phthisis or imbecility in the parents or earlier ancestors, and like conditions in collateral branches of the
family. It is only in exceptional cases that we can
expect a reappearance of the identical phenomena of
degeneration in each new generation, and that of deaf-
mutism is not one of these.

It must not, however, be thought that it is at all
uncommon to find this particular outward sign of
family decay transmitted unchanged from parent to
child, or appearing in several members of the same
family. Occasionally a whole family is found deaf
and dumb. Ribot, who argues strongly in favour of
the hereditary character of deaf-mutism, found that

of 148 pupils in the Deaf and Dumb Institution of London, there was one in whose family there were five deaf-mutes, one in whose family there were four, eleven in each of whose families there were three, and nineteen who each had a brother or sister similarly afflicted with themselves. These figures Ribot considers in themselves conclusive proof of the frequent hereditary character of the affection; and certainly when we find in thirty-two out of 148 cases, almost 22 per cent.—this evidence of the condition not being confined to the individual, but present in other members of the family—the case in favour of the hereditary origin of the affection is strong. But much more positive evidence of the frequent hereditary transmission of deaf-mutism is to be found in the article on " Vital Statistics " in the Report of the Irish Census Commissioners. This writer discusses at considerable length the subject of congenital deaf-mutism, and produces a mass of evidence which Sir William Turner, in his address before the Anthropological Section of the British Association in 1889, asserted "proves that it is often hereditarily transmitted."

This writer states that "in the Irish census for 1871, 3297 persons were returned as deaf-mutes, and in 393 cases the previous or collateral branches of the family were also mute. In 211 of these the condition was transmitted through the father, in 182 through the mother." Here we have almost exactly 12 per cent. of direct transmission through one or other parent. But this writer does not stop here; he gives other important figures. He says:—"In 379 instances there were 2 deaf-mutes in the family, in 191 families 3, in 53 families 4, in 21 families 5, in 5 families 6, and in each of 2 families no fewer than 7 deaf-mutes were born to the same parents."

Now, it is not too much to infer, when a degenerate condition is met with in from 2 to 7 children of a family, that the cause, whatever it may be, comes from the parents, in other words, is hereditary. Here, then, out of 3297 cases we have 651, or 20 per cent., in which the condition is proved to be due to parental defect, from the fact of its attacking from 2 to 7 of the children of those parents. It will be noticed that these figures closely agree with those of Ribot, who found 2 to 5 children afflicted with deaf-mutism in the families of 22 per cent. of the pupils of the London Deaf and Dumb Institution.

How any one with all this evidence before him can come to the conclusion that deaf-mutism does not depend upon parental peculiarity, but occurs merely as a "sport" or freak of Nature quite beyond explanation, and that its repetition in several of the children of certain parents is but a strange coincidence, it is difficult to imagine.

We have also evidence of congenital deafness being transmitted unchanged through several generations. "Mr. Burton, who has paid great attention to this subject, refers to several families where the deaf-mutism has been transmitted through three successive generations, though in some instances the affection passes over one generation, to reappear in the next. He also relates a case of a family of sixteen persons, eight of whom were born deaf and dumb, and one, at least, of the members of which transmitted the affection to his descendants as far as the third generation." *

Here is the tree of a family whose history is vouched for by the Lunacy Commissioners of Scotland, which shows the transmission of deaf-mutism through four generations :—

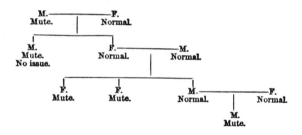

This is a most interesting and instructive case. Besides showing that deaf-mutism may be transmitted directly from parent to child, in spite of the other parent being normal so far as this particular character of degeneration is concerned, it also shows clearly that in many cases where deaf-mutism appears in the children of parents who can both of them hear and speak, the defect in the children is distinctly hereditary. Had we had no family history, it is not to be

* Sir W. Turner's Address to Anthropological Section, British Association, Newcastle, 1889.

doubted that the deaf-mutes occurring in the last two generations of this family would have been set down to some inexplicable freak of Nature. The parents in each case were both of them normal, yet they brought forth deaf and dumb children. A single glance at the family tree shows from whence came the imperfection.

Unfortunately, in the histories of such families as the above, we rarely know anything of the in-coming so-called normal parents beyond the fact that they can hear and speak. We do not know whether they approach the normal type, or are members of families as degenerate as those into which they marry. But in such cases as that before us, in which deaf-mutism or other defect appears generation after generation in the children of normal members of the tainted family, we are justified in assuming that the in-coming parents are also tainted with degeneration, and that it is the double parental taint which ensures the reappearance of the pathological character. For instance, in the family under review the deaf-mutism was only sufficiently potent in the great-grandfather to reproduce itself in one of his two children, yet we find it appearing in his children's children, and in their children again. We have already seen how readily reversion to the normal takes place on the individual bearing the lately acquired character "crossing" with the normal. All through Nature this is the rule, and when we find a character already latent in the individual for want of prepotence reappearing in the offspring of that individual, we are justified in **assuming** that some tendency to degeneration in the other parent is the exciting cause of its reappearance. Therefore we must suspect the in-coming parents who joined the above family in its second and third generations of bearing some degenerate taint; for had they been free from taint, the deaf-mutism would have been absent in the third, or certainly in the fourth generation. The disease tendency in these in-coming parents was not necessarily deaf-mutism. It may have been scrofula, epilepsy, insanity, hereditary syphilis, or any other depraved condition. All degenerate characters are allied, and their promiscuous mingling is slightly, if at all, less dangerous to the offspring than the intermarriage of any one of them.

Sir William Turner says:—"There can be little doubt that congenital deaf-mutism, in the great majority of instances, is associated with a defective development;" * and I think every one who studies the subject will be of this opinion. Every person who has visited an Institution for the deaf and dumb must have been struck with the more or less imbecile aspect of the unfortunate inmates. In the majority of deaf-mutes, the whole economy, mental and bodily, is more or less blighted. Many are scrofulous, and a vast number succumb to phthisis. Physically they are poorly developed; they have round backs and narrow

* Sir William Turner's Address at British Association, **Newcastle,** 1889.

shoulders, and their limbs are badly formed, while the shape of the head and the cast of the features point strongly to a degenerate type.

Physical deformities, such as large projecting ears,* slobbering mouths, squint, paralysis, blindness, and total idiocy, are very common among the congenital deaf, and show the close relationship between deaf-mutism and other forms of degeneration. Professor A. Graham Bell of Washington has made careful inquiry into this particular branch of the question, with the result we might expect. He found that the ratio of blindness among deaf-mutes is $14\frac{1}{2}$ times as great as among the whole population; while idiocy, the deepest of all forms of degeneration, is 43 times as great among these unfortunates as among the general population.

This investigator (Professor Bell) has recently called attention to the fact that between 1870 and 1880 the deaf and dumb population of the United States of America had increased from 10,000 to 34,000. This he attributed largely to hereditary transmission, and he animadverted severely upon the mistaken views of would-be philanthropists, who, not content with mitigating the unhappy lot of these sufferers by educating them and enabling them to do something toward earning a living, hold out inducements to them to intermarry. Unfortunately, America is not the

---

* Albertatti found this character in 50 per cent. of deaf-mutes. It is a commonly recurring character among idiots and instinctive criminals. "The Criminal," Havelock Ellis, p. 80.

12

only country in which such marriages are striven
after and hailed with loud approval. Too many of the
lady patrons of our own asylums for the crippled and
deformed in mind and body, from some mistaken
notion that even in the case of a dumb man or blind
woman a single life must necessarily be incomplete,
depraved, and sinful, are only too ready to encourage
marriage among those who themselves require the
assistance of others to enable them to live. These
ladies should understand that all such marriages are
outrages against Nature's benign laws, and that their
promoters must be suspected of the same morbid
feeling which fills a church to witness the wedding
of a Tom Thumb or some monster from one of Barnum's
side-shows.

Professor Bell says:—" Philanthropy in this country
[America] is doing everything possible to encourage
marriage among deaf-mutes. . . . Unless this system
of management is changed, we shall certainly have
a deaf variety of the human race." But of this I can
see no danger. By encouraging these unhappy creatures to marry and inter-marry, we certainly can increase the numbers of the idiotic, blind, deaf, insane,
scrofulous, deformed, and otherwise degenerate, and
so increase vastly the already too great amount of
human suffering. But few such families will long
survive, and none of them will live in posterity.
Were the absence of the sense of hearing the only
fault in the stock from which our deaf-mutes spring,
there is no reason why, under the artificial conditions

of civilised life, a deaf variety of mankind should not be founded, as Sir William Lawrence said we might found a six-fingered variety. But congenital deafness is not the simple absence of one of the senses which is not absolutely necessary for life; it is a sign of a general decay, which, if deepened by intermarriage, must soon reach the necessarily fatal type and extinguish the family.*

Congenital deaf-mutism may, and does, occur at times in families in which it is impossible to discover any signs of decay of the stock. But even here its appearance is generally to be explained. In such cases, the blight most frequently depends upon initial heredity, for the deaf-mute, like the idiot, is often the result of drunkenness on the part of the parent when the child was begotten. In these cases the deafness is generally accompanied by considerable imbecility or by some physical deformity. Another common cause of deaf-mutism is habitual drunkenness in the parent; but these cases may well be classed with the great mass, for it is always to be suspected that the habitual drunkenness in the parent is itself the result of an inherited taint, which is transmuted in passing to the child.

Another cause of deaf-mutism is syphilis in the parent. This disease, as we have seen, often brings

* Menckel has published cases showing how frequently the children of deaf-mutes exhibit imbecility, *i.e.*, a stage further on the downward course.

about an impoverished, devitalised condition of the system, which is reproduced in the offspring as idiocy, scrofula, epilepsy, deaf-mutism, or some such blight.

Seeing, then, that this terrible affliction is rarely, as some would have us believe, an inexplicable freak of Nature, but in most cases has its origin in some parental character more or less degenerate, it behoves us to consider seriously the advisability of the marriage of those so afflicted, and of their kindred.

Firstly, it should be impressed upon all, and especially upon a certain class of philanthropist, that the congenital deaf should not be induced, or even permitted, to intermarry. The *vis medicatrix naturæ* should here, as elsewhere, be given a chance of improving the family stock, and this cannot be where both parents are tainted.

In the next place, it is doubtful whether deaf-mutes should marry at all. However, if a deaf-mute can induce some normal person of the opposite sex to take him for better for worse—and such a union should have its advantages—we could hardly expect him to say " No " in the interests of those still unborn. But whether any person of reasonably sound development could be induced to enter into such a union is extremely problematical, and in this fact lies the danger. If the perfectly sound could be mated with the deaf-mute, there can be little doubt that in many cases the degenerate character would be lost by the reversion of the children to the normal type. But in few instances will the sound and healthy be fascinated by such unfinished creatures as the deaf-mute, "the poor creatures . . . are specimens of Nature's workmanship in its most untidy mood; features have rarely been duly chiselled; the sense of beauty has long been dead, while *gauche* figures and manners often render the victims little short of repulsive to all but those who, born under like conditions, have their faculties of perception so maimed and blunted that they know no better, and are also debarred from making any higher choice." * Thus, the deaf-mute is thrown back from the society of the more normal to that of his own grade of development, and reversion in his children is denied an opportunity.

As to marriage with the near relatives of the deaf-mute, the wise man or woman will think twice, or even thrice, before entering into such an alliance. Just as the idiot or epileptic child in a family points to a family tendency to degeneration which may take on some other form in the next generation, so the family in which the deaf-mute has appeared is to be looked upon with grave suspicion, or, better still, avoided altogether. The man or woman who espouses the brother or sister of the deaf-mute must not be surprised if blindness, idiocy, or some other terrible imperfection, appear in some of the children.

Fortunately, it is not necessary to put a premium upon the procreation of children in this country. According to some economists, our population increases almost

* J. Russell Reynolds, M.D., F.R.S., address before Congress of the Sanitary Institute of Great Britain, 1887.

too rapidly as it is. Why, then, should the deformed and unfinished be tempted to add to the struggling throng creatures who must sink by the wayside in the fierce struggle for existence? Rather let them be taught the beauty of self-denial, and be induced to refrain from entailing upon others the afflictions which have fallen upon themselves from no fault of their own.

CHAPTER XIII.

CANCER.

"The more I have seen of cancer as occurring among families where the family history is known, the greater becomes the number of cases in which well-marked inheritance can be traced. . . . Now I can without difficulty count as actual facts not less than one in three of the patients with cancer in whose families the occurrence of cancer is well known."—SIR JAMES PAGET.[*]

"We cannot over-estimate the importance of inheritance in the origination of cancer."—SIR WILLIAM AITKEN.[†]

CANCER is at once the most painful, revolting, and fatal of diseases. It spares neither age nor sex, neither rich nor poor. The emperor in his palace and the pauper in his crib are equally within its reach. Once it has marked a person for its own, it clings relentlessly to its victim until the grave closes over him; and on the journey thither, which is sometimes long, it too often racks him with excruciating pain, and renders him such a pitiable and loathsome object, that his friends pray in secret for his emancipation at the hand of death.

Cancer is a disease for which there is no cure known. The knife of the surgeon is our only hope. In cases judiciously chosen, early extirpation, where that is

[*] Pathological Soc. Trans., 1874, vol. xxv. p. 317.
[†] Op. cit.

possible, may give prolonged relief; indeed, death
may occur from some other cause before the cancer
returns. But in the vast majority of cases it does
return—in some cases very rapidly, and seemingly with
increased virulence—and it is generally thought that
if life were sufficiently prolonged it would reappear
in all cases. Sir James Paget records a case which
shows how much and how little the surgeon can do
against the disease. It was the case of a lady " whose
breast he removed when she was five months advanced
in pregnancy. She recovered well from the operation,
and the benefit procured by its performance was very
great. She bore her child, and was able to suckle it
for a year before she died, with her most anxious wish
fulfilled, in comparative comfort." What a prospect
for the unfortunate child !

Of these malignant growths there are several
varieties: the hard, called scirrhus, which generally
attacks persons in the decline of life, and is rarely
seen in persons under forty years of age; the soft or
medullary, which is of more rapid growth, and consequently more rapidly fatal, and which in most cases
attacks the young and those in the prime of life; the
epithelial, which is almost always found in connection
with the skin or a mucous surface; the melanotic,
which is so called from the strange fact that many
of its cells are so loaded with pigment granules as
to render the tissue quite dark in colour; and the
osteoid, in which the malignant growth is made up
almost wholly of bone. All these forms differ more

or less as to mode of onset, rapidity of growth, period of life at which they appear, and tissue or organ forming their favourite seat of attack; but all have one character in common, and that is, once established they never relinquish their victim.

Considering that this terrible disease has been increasing of late at an alarming rate, it is of the utmost importance that people should have some idea of the conditions which appear to give rise to it, more especially if by the dissemination of such knowledge it be deemed possible to limit, in however small a degree, its ravages in society.

It is several years now since it was first whispered that cancer was on the increase. This uncomfortable suspicion gradually spread, and soon infected the general public and the medical profession alike. Proof was added to proof, and what was but a suspicion soon became an established fact. It became evident to all that this cruel scourge of mankind was increasing among the inhabitants of this and other civilised countries. The number of deaths attributed to this disease in the Registrar-General's returns grew steadily —grew out of all proportion to the increase of population; but being a disagreeable subject, it was, like other disagreeable matters, allowed to lie in the background for a time. This, however, as we might expect, had no effect upon the figures of the Registrar-General, and, as these went on increasing annually, it at length became necessary to attack the distasteful subject.

As is usual in all cases where the desire to prove

a negative is strong, an attempt was made to explain away the growing figures of the Registrar-General. This, unfortunately, was found impossible. Then it was argued, on the one hand, that there had always been the present amount of cancer among the people, but that, from want of care and knowledge, it had not been recognised and registered as such; while some, on the other hand, asserted that the recent advance of scientific knowledge in medicine had so excited the judgment of the younger among medical men, that, in an over-anxiety to display their critical discrimination of abnormal tissues, they had been induced to believe malignant what were in reality benign growths. These arguments answered each other, and as no theory was found sufficient to combat the Registrar-General's hard facts, it was eventually admitted that, from whatever cause, this terrible disease was really increasing in every grade of society. The increase was perhaps most marked among the inhabitants of our great centres of population, but that there was an increase, that it was growing steadily, and that it was general, was at length reluctantly admitted.

Sir Spencer Wells, in his Morton Lectures delivered before the Royal College of Surgeons of England in November 1888, was unable to bring forward any explanation of the increase in the death-rate from cancer satisfactory to himself, and openly acknowledged his acceptance of the belief that cancer was rapidly increasing among our people. In his very excellent lectures Sir Spencer brought forward a more elaborate

and carefully prepared array of figures than had ever
before been presented on the subject, and the whole
drift of his statistics was to confirm the generally
accepted belief that the disease was increasing in prevalence, not only among the inhabitants of England
and Wales, but also among those of Scotland, Ireland,
and the United States of America; in fact, among all
civilised peoples from whom statistics on the subject
were obtainable.

The number of deaths from cancer in England and
Wales increased from 7245 in 1861 to 18,654 in 1889;
but as the population increased largely during this
period, these figures do not give a correct idea of the
prevalence of the disease in proportion to the population. The Registrar-General, however, supplies figures
which do convey a correct idea of the prevalence of
the disease; they are these:—In 1861 the number of
deaths from cancer to one million persons living was
360, which number had in 1888 reached 610 by an
increase of marvellous regularity. Of this increase Sir
Spencer Wells says:—"I think it hardly possible that
this steady increase in twenty-six years from 360 to
610 deaths from cancer among each million persons in
England and Wales during that period can be truly
attributed to any great extent to better registration;"
and with this I think all must agree.

Nor, as I have already hinted, is this alarming state
of things confined to England and Wales. In Scotland
deaths from cancer are at present even more plentiful
in proportion to the population than in England; while

Ireland, though considerably less afflicted with the disease than other parts of the kingdom, also shows a steady increase during the past ten years. According to Dr. Fordyce Barker, the condition of affairs is almost the same in America as with us in the United Kingdom. The mortality from cancer had increased in New York from 400 to the million in 1875 to 530 to the million in 1885. In commenting upon this fact Dr. Barker remarks, that the disease is much less frequent among the negroes than the whites, and that the mortality from cancerous disease is largest " in those nations which are the most advanced in civilisation," or, in other words, in those peoples who interfere most with the laws which govern natural life.

And now as to the cause of this disease and of its increase among civilised nations. As to the origin of the disease, I may say at once nothing is known. Cancer has been known and feared from very early times, and various theories as to its cause have been advanced from time to time, but none has met with anything like general acceptance. Recently the vegetarian faddists have advanced and strongly maintained the theory that cancer is the result of a diet too largely composed of animal food ; but this theory has nothing whatever to support it, and is at once disposed of by the fact that the death-rate from cancer is actually higher in Scotland, where the diet of the majority is very largely vegetable, than in England, where it is as largely animal. If this theory were true, the disease would have been rife among such peoples as the

American Indians, who lived almost wholly upon the spoils of the chase, whereas we know that such was not the case.

For my part, I may say I look upon cancer as the outward sign of a constitutional degeneration, brought about, like every other degeneration, by interference with the laws of natural life, and very closely related to those other degenerate states which have for their outward signs such diseased conditions as epilepsy, habitual drunkenness, insanity, suicide, and scrofula. That this is so is, I think, proved by the manner in which it increases with what we are pleased to call civilisation, that is, where the interference with the laws of Nature is most marked, and the vitality of the stock most reduced in consequence; by the fact that it is, like every other degeneration, transmitted hereditarily—of which we shall have ample proof later on; and by the still stranger fact that cancer may be, and frequently is, transmuted in transmission to scrofula, suicide, epilepsy, and insanity.

The cancerous diathesis, then, is a peculiarity of constitution, depending upon degeneration, which predisposes the owner to malignant growths of lowly organised tissue called cancer, and which is transmitted to his offspring, where the outward sign of the lack of vitality may differ widely from that in the parent.

In this predisposition we have the real cause of cancer in the individual. How this predisposition was originally built up we cannot at present tell, but doubtless many elements of unhealth combined in its evolution.

Latterly the opinion has been expressed by some that cancer probably depends upon some micro-organism, and this, I believe, will ultimately prove to be true. In 1887, Scheuerlen, of Berlin, announced that he had discovered a bacillus peculiar to cancerous growths; but closer investigation proved fatal to the discovery. Professor Virchow found the same micro-organism growing upon sections of potato which could not have been contaminated from any cancerous tissue. Nevertheless, this appears to be the most promising direction for investigation, and it is not at all improbable that ere long cancer may be found to depend upon the presence of some micro-organism; which discovery, when made, will in no way affect the theory of the hereditary character of the disease.

Predisposition must exist in nearly every case in which cancer occurs. It is by no means unusual for cancer to follow local injuries to tissues or organs. A blow upon the breast will often be found the starting-point of a cancer in woman, and the irritation caused to the lips by the continued use of a clay pipe, or to the tongue by decayed teeth, is frequently given as the cause of the disease. But such injury or irritation can only be the *exciting* cause, else thousands more than are at present would be the victims of cancer. How many thousands receive injuries to the breast, smoke clay pipes, and have their tongues irritated by broken teeth without result? Those who from such causes get cancer are predisposed to it constitutionally. On this point Professor von Esmarch says:—"One has always to come back to the assumption that a certain predisposition is a *necessary* factor. Without this it is impossible to explain how it is that, in the great majority of cases in which irritation exists, cancer does not become developed."

Without this hereditary predisposition not even inoculation will convey the disease. Alibert and other investigators inoculated themselves with cancerous matter, and Harley and Lawrence performed the same operation upon dogs, without effect; which clearly shows that if there be no predisposition, if the soil be not by nature suitable, even the introduction of the bacillus—if there be a bacillus—or of the poison into the system will fail to establish the disease. Just as some are born with a predisposition to consumption, and fall victims to the disease under conditions harmless to those not so predisposed, so others are born with a predisposition to cancer, and in these the malignant disease lights up from injury or irritation, which the healthy undergo with impunity.

On the ground, then, that predisposition accounts for the appearance of cancer in the individual, the alarming increase of the disease among the population is to be attributed to propagation and cultivation of this predisposition. In the spread of this family predisposition there are two agencies at work. First, and by far the most powerful, is hereditary transmission; for just as epilepsy, suicide, and drunkenness are

in certain families handed down from generation to
generation, so in other families is that degenerate constitutional state, whose most frequent outward sign is
cancerous disease, regularly transmitted. The second
agency at work in the production of the predisposition
to cancer is made up of all the degenerating influences
of modern civilised life, which are constantly deteriorating the race, and deepening this and every other
pathological character.

That cancer is a disease which points to degeneration of the stock in which it appears is clear from
the close relationship which exists between it and other
admittedly degenerate conditions, such as scrofula,
suicide, and insanity. Sir William Gull * years ago
pointed out that scrofulous children were very frequently the offspring of parents who had cancer,
or were members of families in which cancer was
common. On the other hand, the frequency with
which cancer attacks members of families showing
the epileptic, suicidal, or insane diathesis is notorious.
Again, just as we have seen that a commingling or
combination of the taints of insanity and epilepsy,
or epilepsy and drunkenness, &c., tends to further
degenerate the offspring to impotent idiocy — the
lowest stage of degeneration compatible with a continuance of life—so we find a combination of this
degenerate condition which predisposes to cancer with
another form of degeneracy will produce offspring
lower in the scale than either parent. Of this a very

* Address before British Medical Association, Oxford, 1883.

good example is given by Dr. B. Ward Richardson, who says:—"The intermarriage of cancer and consumption is a combination specially fraught with danger." He gives the following case:—"A young man of marked cancerous proclivity married a woman whose parents had both died of pulmonary consumption. This married couple had a family of five children, all of whom grew up to adolescence, sustaining at their best but delicate and feeble existences. The first of these children died of a disease allied to cancer, called *lupus;* the second, of simple pulmonary consumption; the third, owing to tubercular deposit in the brain, succumbed from epileptiform convulsions; the fourth, with symptoms of tubercular brain disease, sank from diabetes, the result of the nervous injury; and the last, living longer than any of the rest, viz., to thirty-six years, died of cancer. The parents in this instance survived three of the children, but they both died comparatively early in life—the father from cancerous disease of the liver, the mother from heart disease and bronchitis." * In this case there was no chance of reversion to the healthy type, and the result was the same as in several of the families already mentioned in the chapters on insanity and epilepsy where both parents were tainted, viz., extinction of the family. Only one of these wretched children lived to die of cancer; but it cannot be doubted that had the others survived longer, more of them would have developed malignant disease.

* "Diseases of Modern Life."

13

As it was, four of them had not sufficient vitality to enable them to hold on to life until that age at which cancer most commonly appears, and they succumbed one after another to tubercular disease, the scourge of those bankrupt in vitality.

Now, in the first place, this "young man of marked cancerous proclivity" should never have married, and it is possible that had he been convinced as to what would be the outcome of his marriage he would have foregone the pleasures of matrimony. But if he *would* marry, then he should have chosen a partner free, so far as possible, from degeneration. Had he done so, it is possible some of the children might have escaped; but as it was, the offspring had no chance of coming back to the path of health, and Nature stamped them out as unfit.

The family history which I gave at page 49 shows very clearly the close kinship existing between the cancerous diathesis and those other forms of constitutional degeneration whose outward symptoms are infantile convulsions, suicide, epilepsy, insanity, tubercular disease, and sterility. The father of this family died of cancer of the stomach at sixty-six years of age. He had a brother who cut his throat at fifty-six; the mother, an apparently healthy woman, "died of a fit" at the age of fifty-four. To this pair seven children were born, as follows:—1. A son who died of cancer of the stomach at fifty-eight. 2. A son who died in convulsions, aged thirteen weeks. 3, 4, and 5. Three daughters who died of consumption, one at the age of sixteen, the other two later in life, and after being married for many years; neither left any issue. 6. A son who is epileptic, and has twice been confined in lunatic asylums; married—no issue; and 7. A son who is up till now sane, and enjoying fair health.

Here the taint in the mother appears to have been slight, still it was there, and while certainly preventing reversion, it doubtless deepened the degeneration of the father in the children. In the father's stock the taint was much deeper, and it is to be noticed that while it was exhibited as cancer in him, it took the form of suicidal impulse in his brother. In the children of this pair we have the disease of the father transmitted to the eldest son, but will any one refuse to believe that the infantile convulsions, the liability to tubercular disease, the epilepsy, the insanity, and the marked sterility, were but the varying symptoms of the degenerate nature, inherited from a father who might have died of some acute disease at any age under sixty-six, taking the secret of his nature with him, and leaving the origin of his children's unfitness a mystery.

Another and a very strong proof that the cancerous diathesis is a family degeneration is to be found in the fact that, when deeply marked, it is not unfrequently accompanied by sterility, just as are the lower grades of the scrofulous, insane, and other degenerate temperaments. On this ground would I account for the fact that cancer so frequently attacks the childless woman. The frequency with which malignant disease attacks the womb, breast, and other organs in this class is generally set down as the result of an abnormal state of those organs, because of their never having been called upon to perform the functions for which they were designed; but I would argue that the degenerate state of the system, which ultimately shows itself in cancerous growth, brought about the barrenness, rather than that the absence of physiological activity was the cause of the disease. In support of this theory, which I believe has never before been advanced, I would point out that a large number of the children of cancerous parents, who themselves may never develop cancer, are childless. We have an example of this in the family whose history I have just given, in which no less than two married daughters and one married son were childless. Further, I would point out that cancer does not attack the unmarried woman, in whom the functions of the generative organs are in total abeyance, to anything like the same extent that it does the barren married woman, in whom the organs are submitted in a great degree to the nervous excitement necessary for their functional health. The children of the cancerous are undoubtedly deficient in vitality, and the deficiency may make itself evident in barrenness, as it may in idiocy, or scrofula, or epilepsy, or insanity. In my opinion the barren woman who develops cancer, or who is the daughter of a markedly cancerous stock, is barren from the same cause that the female imbecile and the prostitute are barren, viz., because she has reached a state of degeneracy at which Nature refuses to continue the race.

Unfortunately cancer is one of those diseases which in the vast majority of cases does not make itself known until late in life, and many who bear within them, and convey to their children, the tendency to this disease, die of some other affection, without ever becoming aware of the curse they have borne about with them through life, and handed on to their children. To this fact is to be attributed those numerous cases where we find several children, of parents who have died without displaying a sign of cancer, dying one after another of malignant disease. On this subject Sir James Paget remarks:—"Now I can without difficulty count as actual facts not less than one in three of the patients with cancer in whose families the occurrence of cancer is well known. But this number does not nearly represent what we may very safely assume to be the predominance of inheritance of cancer. A large number of persons die of internal cancer, and convey it to their offspring, though it is never known that they themselves have been the subjects of cancer, or, at least, is never recorded. A large number more die before they have manifested the cancerous disposition which is in themselves; for, paradoxical as it may seem, if a man have not outlived the utmost age of man, we are bound not to believe that he might not have been the

subject of cancer; for cancer is eminently a disease
of degeneracy, a disease of which the frequency increases as years increase—*i.e.*, in proportion to the
number of persons living at each period of life, the
number of cancer cases increases as age increases.   So
that unless a man have lived to the full age of life,
he may have died of some other disease than cancer,
and never have manifested the cancerous tendency
which he yet conveys *in predisposition* to his offspring.
The cases are very far from rare in which offspring
die of cancer long before their parents.   The parent
lives, and maintains that cancer was never known in
the family, but a few years elapse, and then the
parent dies of the very same disease as the offspring
died of, having been quite ignorant of the conveyance of the disease of which the offspring died." *

But it is not necessary to labour the point.   That
cancer is eminently an hereditary disease is admitted
by all the greatest authorities in the medical world;
and that being admitted, it is impossible to deny
that hereditary transmission must be responsible for
a large proportion of the alarming increase which
has taken place in recent years.   In almost every
case which occurs, if we can only trace back through
three or four generations, we are sure to discover
the taint.   Like every other degeneration, temperament, or diathesis, it is the work of some considerable time to acquire the cancerous diathesis, and I do
not believe it is ever acquired during one lifetime.

* Trans. Path. Soc., vol. xxv. 1874.

Certainly our present artificial mode of life saps the vitality of the race very rapidly, and the more it becomes estranged from natural life, the more rapidly will degenerations be acquired, but that, at the present day, liability to cancer can be built up in healthy stock in one generation, we have no proof whatever.

The last two cases of cancer I had under my care were women (who are much more liable to the disease than men). One was an imbecile, a poor creature with squint, and evidently of a degenerate stock. At first no history could be got, but when she was dying a relative came to visit her, and on being told the state of affairs, said, "Ah, poor soul! her mother died of just the same thing." The other was a single woman, of about forty-four years of age, several of whose brothers and sisters had died of consumption. This was the only family history which could at first be got, but later it was discovered that her mother and mother's mother had died of cancer of the womb. These cases I only mention as being the last I have seen; taken alone, they would prove little. It is quite possible that the occurrence of the same disease in mothers and daughters and grand-daughter was mere coincidence; such, however, could hardly be the explanation of the following case, which I quote from Sir James Paget. A lady died of cancer of the stomach; of her children, one daughter died of cancer of the stomach, and another of cancer of the breast. Of her grandchildren, two died of cancer of the breast, two of cancer of the uterus, one of cancer of the bladder, one of cancer of the axillary glands, one of cancer of the stomach, and one of cancer of the rectum. It is to be hoped that this was the last of this wretched family.

Such cases as the above, where the disease is transmitted unchanged through several generations, are comparatively rare, yet these are the only cases which are at present counted as hereditary. When the transmutability of cancer with other signs of degeneration of the family comes to be more clearly understood, I have no doubt that in almost every case we shall be able to trace the family taint, and raise the percentage of cases depending upon heredity to close upon 100 per cent.

And now, in conclusion, supposing the theory of the hereditary nature of cancer fully accepted, what lessons are to be drawn from our teaching? In the first place, it behoves all those who are aware of cancer in ancestor, or other near blood relative, to avoid everything which might act as an exciting cause of the disease. They must endeavour to keep their health at as high a level as possible, by obeying all the hygienic laws laid down for general observance. They must beware of irritations and injuries likely to light up the sleeping predisposition within them. It would be as wise for the son or daughter of a person who had died of cancer to smoke persistently at a rough clay pipe, or go in the way of getting the breast bruised, as it would for the son of a madman to indulge freely in alcohol. By thus attending to the laws of health and avoiding injuries and irritations of organs, their chances of living to die of something else than cancer will be greatly increased, of this they may be certain, while their pleasure and usefulness in life will assuredly not be lessened.

On the question of marriage it is difficult to give advice. The thoughtful man or woman will hesitate to take for a partner a member of a family in which it is known cancer has occurred—the careless and the sordid will marry as of yore.

I think no one whose parent or grandparent has had cancer—more especially if the ancestor be of the same sex—should marry without much thought for the possible consequences. But if, after consideration, they determine to risk all, then let them choose as partners persons whose family histories are good; persons in whose families neither cancer, scrofula, epilepsy, drunkenness, gout, idiocy, nor insanity is known, remembering that in doing so they are lessening the sorrows and sufferings of their offspring.

If some such suggestions as these were carried out for a period of time equal to two generations, I am convinced that the amount of suffering from this disease would be changed from what it now is to a waning quantity.

CHAPTER XIV.

TUBERCULAR DISEASE.

TUBERCULAR disease, the most common form of which is that known as phthisis or consumption, has been a perfect scourge to the human race from the earliest times. It is to be found in every climate and in all nations, and is at the present day decimating every civilised community on the face of the earth.

Much has been done in England within the past quarter of a century to limit the ravages of these diseases; yet, notwithstanding all that sanitary science has yet accomplished, tubercular diseases were responsible for no less than 64,235 deaths in England and Wales during the year 1889. This is equal to 2213 to the million of persons living, and is actually an eighth part of the whole of the deaths recorded during the year.

As I have said, tubercular disease has been known from the earliest times. Hippocrates described it as attacking the lungs—the most common seat of the disease among the inhabitants of these countries at the present day—but it was not until the twelfth century that the scrofulous diathesis was recognised and described by Gordonius. From that time our knowledge of tubercular disease has been slowly growing; for centuries, progress, if there were any, was very small, but within the past hundred years our knowledge has grown apace, investigation having been rewarded time after time by discovery. It had long been known that tubercular disease could be communicated by inoculation of tubercular matter (as the sputum of a phthisical person), or by feeding an animal upon the flesh of animals which were tubercular; but it was not until 1882 that the disease was robbed of its last secret by the discovery, by Professor Koch of Berlin, of the micro-organism upon which all tubercular disease depends. This eminent scientist discovered that in all tubercular growths there are to be found myriads of micro-organisms having characters peculiar to themselves. This disease germ, which Koch named the "tubercle bacillus," is to be found in every case of tubercular disease, and will, if introduced into the system of an apparently healthy animal, light up tubercular disease. It can, moreover, with proper attention to temperature, and in a proper medium, be multiplied indefinitely outside the living body altogether, and such artificially cultivated germs, when inoculated upon the living animal, produce tubercular disease, just as would matter taken from a diseased person or animal.

Professor Koch's discovery of the tubercle bacillus, although the existence of the germ had for some time been suspected, caused great stir in the medical

world, and had the effect of greatly unsettling opinion
upon tubercular disease generally.   This was to be
expected, for what had been looked upon for ages as
one of the most hereditary of all diseases, was suddenly
proved to be a contagious disease, which could be conveyed from person to person, like small-pox or scarlet
fever, and doubt at once arose as to whether hereditary
predisposition had anything whatever to do with the
spread of the disease.   This doubt was fostered by
the discovery that the sources of infection were almost
innumerable.   It was shown that the disease might
be conveyed by eating the flesh of animals suffering
from the disease, or by drinking the milk of such;
and as it was known that a great number of cattle
were tubercular, it was evident that here alone was
a great and constant source of contagion.   It was
further pointed out that the sputa of persons suffering
from phthisis contain myriads of bacilli, and as such
persons are constantly coughing and spitting, there
must be grave danger to all who live with, work
with, or have intimate relations with, such persons.
Nor did the danger of infection from phthisical
persons stop even here, for it was asserted that this
bacillus-laden sputum, after it had become hard and
dry, might be broken up, to float as minute particles
in the air, and be inhaled by any one breathing the
contaminated atmosphere, the bacilli in such dried
sputum retaining sufficient vitality to set up tubercular disease in the person inhaling it.   In fact, the
disease germ was almost ubiquitous, which fact alone

was sufficient to account for the prevalence of the disease. Indeed the wonder was, not that so many suffered, but that more did not suffer from a disease, the germs of which were to be found on every hand. This was the position taken up by some immediately after Koch's discovery, but already opinion on the subject is toning down, and at present there are very few who do not admit the existence of a predisposition to tubercular disease, which is hereditarily transmitted.

It has been proved conclusively that tubercular disease can be conveyed from individual to individual, apparently regardless of temperament or diathesis, by the introduction of the tubercle bacillus into the system, but that there exists a diathesis which predisposes the owner to the attack of this particular disease germ, there can be no possible doubt. That this peculiar constitutional state is a degeneration, that it is, like every other degeneration, hereditary. and that it is frequently associated, both in individual and family, with other degenerate conditions, such as idiocy, insanity, deaf-mutism, cancer, drunkenness, epilepsy, and crime, it is now my business to prove.

It is impossible to guess to what this great discovery of Koch's may lead in the near future. Since the bacillus can be cultivated outside the body in artificial media, there is no reason why, by varying the conditions under which it is so cultivated, the germ itself should not be altered in character and rendered less virulent. This might lead to protective inoculation like that practised against small-pox. Nor is there, seemingly, any reason why some other micro-organism or chemical compound should not be discovered which, itself innocuous to the animal organism, would materially modify the virulence of the tubercle bacillus in the system, or destroy the noxious germ altogether. Indeed it might be said that "all things are possible" in this department of medical science. At present hundreds of workers all over the world are labouring earnestly in this field of study, and are adding almost daily to our knowledge of these seeds of disease. Science can never eradicate disease, but it can prune it and keep it within reasonable bounds; and the discovery of the micro-organisms, upon the presence of which so many diseases have already been proved to depend, will vastly aid Science in this work.

Let us now briefly consider the diathesis of which we have spoken, which predisposes so strongly to the attack of tubercular disease.

As there are two distinct types included in the tubercular diathesis, it will simplify matters to take each type separately. I shall therefore first give a very brief enumeration of the characters typical of what we shall call the Phthisical Diathesis, and afterwards a short description of the true Scrofulous Diathesis.

What we have called the phthisical diathesis is generally marked by the presence of a clear complexion, a fine skin, and features well cut and often beautiful. The lips are red, the teeth pearly white, though liable to early decay, and the eyes are large and full, the pupils being widely dilated and the white of the eye beautifully clear. The eyelashes are long, curved, and silky, and the blue veins show distinctly through the clear thin skin; the bones are light, the hands and feet well formed, the stature often tall, and the whole figure slightly and gracefully built. Persons of this type generally remain spare, and they have a strong dislike to every kind of fatty food. They are vivacious and excitable, and the intellectual faculties are often highly developed. Even at an early age children of this temperament in many cases show a marvellous intellectual activity, and it is observation of the regularity with which such precocious tubercular children die that has given rise to the common saying, when speaking of exceptionally clever children, that they may be " too wise to live long."

These persons are wanting in stamina in the widest sense of the term. They are incapable of prolonged exertion either of mind or body, and break down under conditions which would not prove injurious to the healthy. They are continually taking " colds," and are specially prone all through life to affections of an inflammatory character.

Although large families are often born to parents of this type, the children are deficient in vital force, and are carried off in such numbers during infancy by convulsions, brain fever, water on the brain, exhausting diarrhœa, and other ailments, that only a small proportion of those born ever arrive at maturity, and few indeed reach old age. Sooner or later in life the majority of these persons develop consumption; many are so carried off even in infancy; great numbers succumb before or about the time of adolescence, and only a small remnant live beyond thirty-five or forty years of age.

Persons of this phthisical temperament are also remarkably prone to fatal degenerative changes (fatty degeneration) of certain vital organs, as the liver and kidneys. It is also to be noticed that in them the generative organs are often but poorly developed, which in itself is positive evidence of progressive decay of the stock. It is true that the intellectual faculties are often active and well developed, but even here there are unmistakable signs of the decay which has attacked the system generally. However brilliant intellectually, they are as a class emotional, impressionable, and impulsive, and there is a marked absence of that stability which indicates true mental strength. From slight causes they develop convulsions in infancy, chorea, hysteria, and other nervous disorders in youth, and acute attacks of insanity in adult life.

This diathesis appears to be built up with equal certainty by impure air, drunkenness, and want among the poor, and by dissipation and enervating luxuries among the rich. From either set of causes it is capable of rapid development, and it is transmitted to the offspring with very great certainty. By injudicious marriages and persistent ignoring of the laws of health the necessarily fatal type is soon reached, and to this must be attributed the extinction of hundreds of families every year. In some families, even in the highest ranks of society, the susceptibility to the tubercle bacillus becomes so great, that, despite all modern science backed by wealth can do, the children die one after another in infancy, or succumb on the approach of adolescence. In other cases the degeneration from intermarriage or some other cause becomes more or less mixed in character, and while some of the children succumb to tubercular disease in infancy or later in life, idiocy, suicide, epilepsy, insanity, or the true scrofulous cachexia will appear in others.

This phthisical diathesis might be described as a general degeneration, very closely related to the neurotic, which occurs in families once decidedly above the lower stages of development, but now on the down grade of general decay. Such family decay being the result of the repeated exposure of ancestors to the devitalising attack of the tubercle bacillus, or some other exhausting disease, or to some of the thousand and one evil influences which are constantly at work producing progressive deterioration among all civilised peoples.

Let us now look at the true scrofulous type, which in many, indeed in most points, is the extreme opposite of that we have just been considering. Here the skin is usually thick and sallow or pale and spongy; the features are coarse and ill cut; the eyes dull,

14

and the mouth large. The margins of the nostrils
and the upper lip are frequently swollen, and the
edges of the eyelids, devoid of hairs, often present
a red, raw margin, which greatly disfigures the face.
The whole expression is dull, heavy, and more or less
repulsive, the very antipodes of the quick, eager,
spirituelle, and often beautiful face seen in the phthisical type. Nor does the dissimilarity to type stop
here, for in the scrofulous we find the bones thick
and heavy, with their ends, and consequently the
joints, large. The head is often large and misshapen,
the hands short, the fingers thick, the figure stunted,
with a decided inclination to pot belly. Sometimes
the skin is so loaded with greasy sebaceous matter as
to give it a dirty, scaly look. Discharges from the
eyes, nose, and ears are common, and such discharges
are often most offensive, as is frequently the perspira-
tion of the feet. In such individuals the circulation
is always weak, as shown by the cold hands and feet;
the digestion is generally poor, and they are liable to
"colds" on the slightest exposure. The glands round
the jaws are nearly always more or less enlarged, and
they are liable to the formation of abscesses, which
run a remarkably chronic course. Slight injuries,
which in the healthy would hardly be noticed, in the
scrofulous set up inflammations which often lead to
ulcerations of the soft tissues, destruction of joints,
and disease of the bones themselves, which often continue for years. They are also liable to many of the
most severe and intractable forms of skin disease, in

their case often accompanied by the formation of pus, while they have almost a monopoly of *lupus,* a disease which often eats away considerable portions of the face and other parts, and which appears to stand midway between scrofula and cancer.

Occasionally the mental faculties in the scrofulous are preternaturally developed during early life, but such development is exceptional; as a rule, they are dull, and altogether of a low type intellectually.

It is evident from the foregoing brief and imperfect descriptions that the scrofulous is a very much lower type of degeneration than the phthisical, and, as we would expect, we find it occurring much more commonly in association with other extreme forms of degeneration, such as idiocy, imbecility, and physical deformity.

That the tubercular diathesis, whether of the phthisical or scrofulous type, is a true degeneration, is evident from its hereditary character, the frequency with which it appears associated with other degenerations in the individual, and the perfect interchangeability existing between it and most of the other expressions of decay in the family.

Of this relationship and interchangeability with other signs of family decadence we could not have a much better example than is offered in the family history given at page 186, in the chapter on cancer. In that case the cancerous father and neurotic mother produced highly neurotic and cancerous male children, while the females were so devitalised by the combined parental taints, that all succumbed to tubercular disease. The connection between cancer and tubercular disease was pointed out by Sir William Gull, who called attention to the frequency with which cancerous parents, or parents belonging to families in which cancer was common, begot scrofulous children.

The frequency with which the scrofulous diathesis is met with among idiots is most remarkable. Dr. Ireland, one of the greatest living authorities upon idiocy, estimates that at least two-thirds of all idiots are scrofulous; while Dr. Clouston of the Royal Edinburgh Asylum, who has closely studied the relationship existing between the tubercular diathesis and insanity, treats the matter as beyond dispute, merely remarking that " the frequent association of the depraved nutritive condition known as ' scrofulous' with idiocy and congenital imbecility is well known, and universally recognised by those who have had experience in such cases."

Further proof of the fact that the scrofulous temperament is a deeper level of degeneration than the phthisical, is found in the facts that the phthisical rarely or never develop the purely scrofulous forms of disease, although the scrofulous often develop and die of tubercular disease of the lungs; and that the children of the phthisical are very often scrofulous, whatever other mark of degeneration they may bear. This could not have stronger support than it receives at the hands of Dr. Fletcher Beach, the superintendent of the Asylum for Idiots at Darenth, who, in the course of his extended inquiries into the causes of idiocy, found that phthisis, either alone or in combination with some other degenerate condition, as insanity, epilepsy, or drunkenness, was present in 50 per cent. of the parents of all idiots admitted into his asylum; and as two-thirds of these idiots are themselves scrofulous, it is clear that phthisis in the parent not only deepens to scrofula in the child, but to that lowest of all types of humanity, the scrofulous idiot.

As to the very close relationship existing between insanity and the phthisical type of the tubercular diathesis, there is not room for the smallest doubt. That the phthisical and insane diatheses are interchangeable is proved to the asylum physician every day. Lugol found insanity so common amongst the parents of scrofulous and phthisical persons, that he treated of hereditary scrofula descending from paralytic, epileptic, and insane parents. Schroeder van der Kolk, a quarter of a century ago, showed that phthisis and insanity are interchangeable. So long ago as 1863 Dr. Clouston described a form of insanity peculiar to those of the phthisical diathesis and their descendants, and said he could count hereditary predisposition in 7 per cent. more of such cases than of the insane generally. After twenty-seven years' study and observation he remains of the same opinion. Speaking of phthisis and insanity, he says :—" It is surprising how often both diseases occur in different members of the same family. No physician in extensive practice but has met with many such families." *   On the same point Dr. Maudsley writes :—" There is no question in my mind that insanity and phthisis are met with as concomitant or sequent effects in the course of family decadence." †

Again, it is a matter of common observation that the children of the gouty and the syphilitic are very frequently scrofulous, and to these, and the insane taint, aided by dissipation and enervating luxuries, is to be attributed the appearance of a type so degraded as the scrofulous among the children of aristocratic and even royal families.

Habitual drunkenness in the parents is another fruitful source of the scrofulous temperament, but this has already been shown to be so essentially an expression of the neurotic type, that it is only necessary to mention it here.

Another cause of scrofula in children, and one which gives further proof of the degenerate character of the type, is senility in the parents. It has for ages been popularly believed that the child begotten of the aged father has not the vital energy and recuperative power of the child of the father in his prime ; and that this belief is well founded the recent investigations of Marro, Dr. Langdon Down, Korosi, and others prove conclusively. All observers agree that the senility of the father may, to a great

* *Loc. cit.*   † "Pathology of Mind," p. 112.

extent, be neutralised by the youth and vigour of the mother, but when the mother has passed her youth, the senility of the father is invariably more or less disastrous to the child. It is most wonderful how many idiots, scrofulous and otherwise, instinctive criminals, and drunkards, are found upon inquiry to belong to the class whose mothers were no longer young, and whose fathers were in the decline of life, when they were begotten. This class also show how small is their inheritance of vitality by falling a prey in great numbers to tubercular affections.

Looking upon the instinctive criminal as we do, as the representative of a decaying race, we naturally expect to find tubercular disease actively at work among this class, and in such expectation we are not disappointed. The fact is, the majority of this wretched class die of tubercular and nervous diseases. Here again the relationship of the tubercular and nervous affections is to be noted. The instinctive criminal class belong to the unfit; "they are scrofulous, not seldom deformed, with badly formed angular heads; are stupid, sluggish, deficient in vital energy, and sometimes afflicted with epilepsy. . . . They spring from families in which insanity, epilepsy, or some other neurosis exists, and the diseases from which they suffer and of which they die are chiefly tubercular diseases and diseases of the nervous system." * Mr. Bruce Thomson, Marro, Lombroso, Dr. Wey, and every other observer who has studied

* Maudsley's "Responsibility in Mental Disease."

the criminal, agree that tubercular disease is constantly met with in criminals themselves, and in their ancestors and descendants, and that a majority of the whole total succumb to these diseases. Recently Dr. Pauline Tarnowsky has been very closely studying the prostitute, who may be taken as the analogue of the male instinctive criminal of the petty class, and of 150 women of this class whose family history she was able to get, she found phthisical parentage in no less than 44 per. cent.*

Thus, we see, are the family degenerations all allied, the scrofulous being related to the cancerous, gouty, epileptic, insane, syphilitic, drunken, and criminal, and all these being related to each other.

It must not, however, be inferred from all this that tubercular disease only attacks those inheriting a predisposition thereto. True, the family taint can be traced in 30 to 50 per cent. of all phthisical persons, and in quite as large a proportion of those suffering from scrofulous disease (and that only reckoning taint as existing where tubercular disease has been present in the ancestors), but that inherited predisposition is absolutely necessary to the development of these diseases has been proved to be erroneous. As a rule, the tubercle bacillus only attacks those deficient in vitality, but it is not necessary that this deficiency should be congenital; such vital poverty may be acquired. Of course, individuals thus vitally reduced—brought below par,

* "Étude Anthropométrique sur les Prostituées et les Voleuses."

so to speak—are specially liable to the attack of all other disease germs as well as that of tubercle ; or, if not specially liable, they are, at least, less capable of resistance when they are attacked. Still it must be recognised that the tubercle bacillus has a special affinity for the tissues of those wanting in vital force. Some disease germs appear to attack the robust and the feeble with equal frequency and virulence ; others, as that of typhoid fever, seem actually to prefer the tissues of the physically well-to-do, but the tubercle bacillus has a decided preference for those bankrupt in vital energy from any cause, except perhaps senility.

I would put it in this way :—The tubercle bacillus being almost ubiquitous, so soon as the system reaches a certain level of vital depravity, which I would call "the tubercular level," the individual becomes liable to the attack of the bacillus. In some cases, small-pox, syphilis, or other exhausting disease, leaves the system for a time in an impoverished state, and before the vitality is restored, tubercular disease is set up ; in a host of others, starvation, dirt, habitual drunkenness, and want of fresh air reduce the system to the level of susceptibility with like result ; but in the vast majority who develop tubercular disease, the protective vital level has never been reached, a wretched parentage being unable to confer so much. Such individuals are from the moment of conception helpless against the attack of the tubercle bacillus, and succumb on first contact with the germ. We have instances of this wretched class in the scrofulous idiot and in the children of those families where the offspring die off one after another, soon after birth, of various tubercular affections.

The sources of infection by the tubercle bacillus are almost innumerable. It may be inhaled with matter floating in the air, or be introduced through any cut, scratch, or other breach in the skin or mucous surface, or it may be taken into the system with food. Cattle are specially liable to tubercular disease. Herr von Gossler, in his speech before the Prussian Diet, stated that 10 per cent. of all horned cattle slaughtered for food are tuberculous. This estimate may be high, but it is certain that a vast number of the animals are tuberculous, and that the consumption of the flesh of these diseased animals is one of the greatest dangers to which the human species is exposed. The milk of such cattle often swarms with bacilli, and the use of this, or of butter made from such milk, is dangerous in the extreme.

To lessen as far as possible the risk of infection, the ventilation of all rooms in which either persons or animals live or work should be strictly attended to. Living or sleeping in the same room with one suffering from tubercle should never be permitted. The sputa of phthisical individuals should not be ejected here, there, and everywhere, but received in a vessel containing some strong disinfectant, and be burnt or buried afterwards. Cattle kept for dairy purposes should be regularly examined, and those found tainted with tubercular disease at once destroyed. All meat offered for sale should first be examined by experts. And lastly, all animal food, from whatever source, should be properly and sufficiently cooked. A high temperature kills the bacillus, and the danger of infection from diseased meat might be greatly reduced if all animal food were properly cooked.

The tubercle bacillus attacks most animals whose bodily temperature favours its growth. It is common in beasts and birds, and has even been found in reptiles, but the temperature of these latter is not, under ordinary circumstances, sufficiently high for its growth. As I have said, it is extremely common in the ox, though why it should, has not yet been explained. It is the cause of death in the majority of monkeys, elephants, lions, tigers, and other wild animals and birds held in captivity. These animals being robbed of their natural exercise in the open air, too often huddled together in unhealthy pens or cages, and poorly or improperly fed, often doubtless upon tuberculous flesh, become broken in health, and so devitalised that they fall easy victims to the disease germs, just as man does under like conditions.

In quadrumana the disease runs the same course as in man; but in other animals, mammals and birds, its course is often so very different, that it is only the presence of the micro-organism which proves the identity of the diseased conditions.

It attacks some animals much more frequently than others. Thus it is very common in the ox, and very rare in the horse, a perfect scourge among grain-eating birds, and much rarer among those that eat flesh. Why this predisposition should exist in some animals, we do not at present know, any more than why the negro and the West Indian creole should be specially susceptible to the attack of the bacillus; but it is possible that as our knowledge grows, we may come to understand this too.

As to advice respecting marriage, it may at once be said that those already suffering from any form of tubercular disease should not marry. Neither should any one marry a member of a family in which consumption, or other form of tubercular disease, is common. The clear-skinned, bright-eyed, eager, ethereal creature may charm the eye, and she may be good as she is beautiful, but she can never be the mother of strong and healthy children. It is possible that in the near future Science may be able to eradicate the tubercle bacillus from any individual it has attacked; but even if this feat of Science were accomplished, it is not too much to say, that persons of such low vitality as most of these healed ones would be, would hardly be the kind of partners sensible men and women who wished to live in distant posterity would choose. Until Science can not only eradicate the disease, but instil sufficient vitality into the purified one to prevent a re-invasion of the system by the disease germ, those who are tubercular or have been tubercular cannot be looked upon as favourable candidates for marriage.

On the other hand, all those who, though not them-selves tubercular, are members of " delicate families," families in which tubercular disease has appeared, should, before entering the marriage state, lay their case honestly before their medical adviser, and take his advice. Much can be done for the children of delicate parents by judicious treatment, but to be as effectual as possible, it should be begun with the life of the child, if not earlier. The child can be as effectually treated before birth through the system of the mother, as it can be after birth, and such preven-tive treatment cannot be begun too soon or carried out too rigidly.

CHAPTER XV.

GOUT.

"Gout is one of the most striking examples of hereditary disease, and once established, it may be transmitted for several generations, even when every endeavour is made to eradicate it; but as the contrary is generally the case, the malady being, as a rule, more or less intensified by pernicious habits, it becomes in most cases a permanent legacy."—SIR FREDERICK T. ROBERTS.*

THIS is a disease of great antiquity. As far as we can go back in medical literature, it is one of the diseases which we find described, and some of the earliest of these descriptions come wonderfully near to what we find the disease to-day. Hippocrates, three hundred years before the time of Christ, described this disease with accuracy, and later, Celsus, Galen, Aretæus, Cælius Aurelianus, and many others wrote concerning gout, hitting off its leading characteristics with great fidelity.

Gout is a disease of civilisation. It is one of the degenerate conditions induced by interference with the natural life of the human animal. So long as man remained in the natural state, and gained by physical exertion his living, this disease was unknown—in fact,

* Quain's "Dictionary of Medicine."

it is unknown among the savage peoples even at the present day; but so soon as he entered upon the civilised state, became a chief or a king, and lounged in idleness while others performed that labour which must be done by or for every creature, if the creature is to exist, then this disease appeared; and as civilisation spread, and the non-working class increased, so did gout. In fact, so soon as man began to eat too much and labour too little, gout attacked him. It is above all other diseases the scourge of the opulent and idle.

The vegetarians—those hopeless faddists—have asserted that gout, in common with every other ill that flesh is heir to, is the direct result of animal food. So long, they say, as man remained a vegetarian, gout was unknown. To this we would reply: Yes, and for ages after he had ceased to be a vegetarian. It was only when man became indolent and lazy, on discovering that he could, by exercising his ingenuity, procure more than enough of the best of the wherewithal to sustain life without physical exertion, that the disease appeared. The North American Indians and many other peoples of whom we know were largely, if not entirely flesh-eaters, yet gout among such peoples was unknown. And why? Simply because the very active life they lived in the open air used up all the food stuff taken, and accumulation was impossible. The organs were never overloaded, or, if they were, it was only on occasions which alternated with periods of healthful want, consequently disease from that cause was absent.

Mr. Jonathan Hutchinson has said:—"Had mankind continued to be vegetable feeders, and never known the use of wine or beer, we should have had no experience of gout." But with Mr. Hutchinson, even when he adds the "sole cause" of the total abstinence party to that of the vegetarians, I cannot agree. I can see no reason why by gluttony and indolence the system could not be overcharged with nitrogenous matters from the vegetable world, and, if this condition were maintained for a few generations, why we should not have gout as a consequence. Animal food is more nitrogenous than most, and less nitrogenous than some vegetable products, and I fail to see why the nitrogenous constituents of a vegetable diet should be less injurious than those of an animal diet. So far as we know at present, if taken in equal quantity and equally diluted, the result of animal and vegetable products is much the same. Again, as to wine and beer, both, by the way, vegetable products, it is clear that it is not the alcohol, but rather the sugary matters which are the gout-producers. Our port wines, and Burgundies, and beer, and, most potent of all, our stout, are recognised as our great fluid gout-producers, while whisky, which is a much more concentrated solution of alcohol than any of them, has little if any effect in the evolution of this disease.

The fact of the matter is, the cause of gout is indolence coupled with gluttony. Rich foods are freely partaken of, and sufficient exercise to burn that food off is not taken, hence the organs whose business it is to cast out of the system effete matters have a strain put upon them. This strain sooner or later causes disorder of those organs, which further complicates matters, and soon leads to retention within the system of matters offensive to health. The kidneys are the great blood-cleansers, and these organs are more or less diseased in every case of gout.

Abernethy's receipt for the cure of gout, " Live on sixpence a day and earn it," still holds good, and if the humble coin have been earned by honest, healthy, physical toil, I care not whether it be expended upon steak and kidney-pie or upon potatoes and milk. In neither case will the input of material exceed that needful as a force-producer, and so no accumulation can arise to clog the system. I admit that the glutton may be more likely to overload the system, feeding upon rich dishes of animal food, than feeding upon the less succulent yet no less rich vegetables. But if the rising generation of vegetarian cooks can produce equally seductive dishes with those who go for ingredients to the animal world—as vegetarians boast they can—I fail to see salvation for the gouty gourmand in vegetarianism.*

Gout, then, is essentially a disease of civilisation.

* Since the above was written, Mr. Jonathan Hutchinson, in his "Archives of Surgery," No. I, vol. iii., has forbidden the use of fruit to all patients having a tendency to gout. The contained sugar is, of course, the deleterious agent. The more sugar a fruit contain the more hurtful is it. Cooked fruit eaten with added sugar is specially dangerous.

It is a condition brought about by continued over-feeding and prolonged indolence, and is consequently to a large extent a disease of the wealthy, who can eat as much as they please and work as little. In times past it was almost wholly confined to the rich, but in more recent times some among our working-classes, who have special opportunities for feeding, and whose physical labour is not great in proportion, have cultivated the disease. At present gout is not at all uncommon amongst butchers, bakers, draymen, brewers, innkeepers, coachmen, butlers, publicans, porters, and others who eat and drink largely, and who partake sparingly of physical labour.

As might be expected, temperament has much to do with the cultivation of this disease among a people. As a rule, the large eaters are the less active. They are capable of exerting great power when put to it, but they rarely have any of that restless activity which keeps the thin man thin. They do not care much for the more volatile preparations of alcohol, as whisky, which would induce an uncomfortable restlessness, but prefer beer or other soothing draught. They are generally largely built and have heavy limbs; even early in life they become fat, the skin being oily and the pores in it large. They have great powers of digestion, and to them the pleasures of the table are of the first importance. This type—the sanguine—is the class most given to the cultivation of gout. It is, perhaps, best seen in the inhabitants of the Midland and Southern counties of England, where ages of security and comparative prosperity have made them what they are; or among the Northern Germans and the Dutch, whose pleasures of life are largely made up of eating and drinking.

There is no disease the hereditary character of which is more fully and generally recognised than gout; in many families it is looked upon as an heirloom. Sir Alfred Garrod said he could trace direct heredity in 50 per cent. of all cases; Sir Dyce Duckworth gives 50 to 75 as the percentage of cases he found hereditary; while Sir C. Scudamore (even the medical men who make a speciality of the treatment of gout become aristocratic) traced direct heredity in 60 per cent. of all his cases. Many observers put the influence of family taint at a figure even higher than any of the above, while some have gone so far as to declare the disease purely hereditary (Dr. Cullen). And probably this is true in a certain sense, for although no ancestor may ever have actually had gout, the predisposition may have been building up for some considerable time. Rarely, I believe, is gout built up in a single generation, and when it is, it is not likely to be well developed until late in life, hence little would be conveyed to the children. I think it is certain that well-marked predisposition to gout is in every case the work of several generations.

In many rich families the disease has been handed down through great numbers of generations. A good example is given by Sir Alfred Garrod, who writes:—" A few years since, I was consulted by a gentleman labouring under a severe form of gout with chalk-stones, and although not more than fifty years old, he had suffered from the disease for a long period. On inquiry, I ascertained that *for upwards of four centuries* the eldest son of the family had invariably been afflicted with gout when he came into possession of the family estate." * This fact might be taken as going strongly to disprove my assertion that the predisposition to gout, like every other hereditary pathological character, is a true family degeneration. It might be argued, that if it were a progressive degeneration, the necessarily fatal type must be attained, and the family become extinct, before the lapse of such time as it has been known to run in families like that mentioned above ; and were gout governed by the rule which guides the neurotic, cancerous, scrofulous, and some other family degenerations, this argument would be good. In this, however, gout is peculiar, that it is not nearly so rapidly built up as other family degenerations, and consequently is longer in reaching the fatal type. This slowness in its evolution arises principally from two causes, viz., (1.) The mitigation it suffers during the period of infancy and youth in each generation, and (2.) The difficulty with which the female is affected by this form of degeneration.

Gout is a disease which, except in cases where

* " Gout and Rheumatic Gout," by A. B. Garrod, M.D., F.R.S.

the family predisposition is exceedingly strong, does
not make its appearance until middle life, or even
later. There are instances recorded of gout appearing
even before puberty, but such cases are rare, and
only occur in families whose members have exhibited
the disease for many generations. It is essentially
a cumulative disorder, and the limited, or rather
natural, feeding, and great muscular activity of
infancy and youth in each generation, so to speak
reduces the accumulated poison capital, so that it
is only in very rare cases that it can proclaim itself,
before a personal indulgence in the vicious habits
from which it originally sprang gives it the necessary strength. Sir Spencer Wells pointed out, some
years ago, that the children begotten before any
acute attack of gout in the parent, were but slightly
predisposed to the disease, as compared with those
begotten after the parent had actually suffered an
attack. Now, as few fathers develop gout until
middle life, it is clear that many of their children,
and especially the eldest, must receive the family
taint in a mitigated form ; and as such children are
generally properly clothed and fed, take good and
sufficient outdoor exercise, and are otherwise cared
for hygienically, it is also clear that the weak inherited taint cannot develop much until the vicious
habits—also inherited—become a part of the routine
of life, which seldom or never occurs before twenty
to thirty years of age. Thus, we see, is the gouty
degeneration retarded in its course at every genera-

tion. First, it is inherited in a mitigated form; and second, it is, so to speak, being lived down during the first twenty-five years of life. In the neurotic, the scrofulous, and some other hereditary degenerations, special opportunities for their development are offered during the early years of life, whereas in gout the tendency is materially reduced during that period.

As to the second cause given above of the slow evolution of gout, viz., the difficulty in strongly infecting the female with the predisposition, I would venture to say it is not yet clearly understood. Women certainly do not subject themselves to the same extent to predisposing causes as do men. They indulge less freely in the pleasures of the table, whether of luxuries solid or fluid, and it is rarely that their digestive powers are anything like equal to those of their male relatives. Yet all this will not account for the strange fact of every-day occurrence, that a younger daughter of a family strongly predisposed to gout, and whose brothers, elder and younger, one after another develop the disease, will show no disposition to follow their example. She has inherited the taint, yet she does not develop the disease. So long ago as the time of Hippocrates it was suggested that the catamenial losses experienced by women acted as a safety-valve for the gout poison, and that there is truth in this supposition of the ancients is not doubted in the present day. Unfortunately in these later days woman is fast losing her old-time exemption from this disease. As she apes man in his worst as well as his best habits and customs, so is she acquiring the diseases which were once peculiar to him because of those habits. Now-a-days it is nothing strange to find a woman suffering from gout in its most virulent form, though happily sufficient of the old feminine spirit survives to make such a thing not an every-day occurrence. Sir A. Garrod says, " In the de-generate times of the Roman Empire, when women gave themselves up to every kind of licentiousness, they appear from Seneca's account to have become the subjects of gout equally with men." * What *we* shall ultimately arrive at can only be surmised, but it is perfectly certain that as woman approaches man in education, occupation, and mode of life, she is at the same time acquiring those diseases and defects which were once peculiar to man. Such diseases as gout and general paralysis of the insane, once peculiar to the male, are becoming more common among women every day. Criminality, too, has been increasing steadily among women in England during the past quarter of a century or more. Can any one for an instant doubt the cause ?

Gout, as we have seen, commonly makes its appear-ance between the ages of thirty and fifty years. At times it appears so early as ten, twelve, or sixteen years, but such cases are rare, and in nearly every instance depend upon strong hereditary taint, occurring most frequently in the children of elderly fathers who have suffered repeated attacks of the disease. When

* *Op. cit.*

gout appears in women, in the vast majority of cases it is not until after the "change of life" or old age, and there is almost invariably a hereditary predisposition present.

As to gout being a constitutional degeneration affecting the whole system, there is ample proof. In the first place, almost every tissue in the body of the person who has inherited the gouty diathesis is liable to degenerative change. At an early stage of the disease the kidneys become unfitted for their work because of disease consequent on tissue degeneration. The muscular tissue of the heart as well as its valves is in these persons liable to degenerative change, and the tissues of the blood-vessels, great and small, are more or less diseased in nearly every case. The bones, muscles, connective and cutaneous tissues are also very frequently affected, such persons being liable at all times to inflammations in all parts of the body, to the formation of abscesses and to extensive ulcerations. The disease is not confined to any one, or even two or three, tissues or organs; the whole constitution is deteriorated, every tissue is to a certain extent robbed of its vitality, and the system may be said to be "below par." The reduced or unhealthy condition of the gouty system is shown in the high mortality which attends ordinary acute diseases, and also by the fact that anything which tends to lower still further the nervous tone, such as sexual excesses, severe study, indigestion, shock, lead-poisoning, &c., will at once bring on an attack of the acute symptoms of gout itself.

In fact, anything requiring expenditure of nervous energy and calling upon the stock of reserve vitality finds the account overdrawn, and so causes a temporary or permanent health bankruptcy.

The degenerate condition of the gouty parent is also marked in many instances in his children, more especially those begotten late in life. These not seldom present the scrofulous diathesis more or less deeply marked, or, short of this, they are feeble generally, enjoy only precarious health, and are short-lived.

The old idea that the gouty are specially long-lived, and are also endowed with more than the ordinary quantum of intelligence, is diametrically opposed to the truth. Certainly some of the gouty live to be old, but for every one who does, scores die comparatively young because of their inherited disease tendency; while against every gouty one who holds a proud place in the intellectual world might be placed a crowd of fellow-sufferers made up of coal-heavers, publicans, broken-down butlers and brewer's men, who lay no claim to the intellectual; or even of scions of noble houses once known to fame, but now, alas! only forming that remnant of the decaying aristocracy which the democratic reformer delights in holding up to ridicule.

The following history of a gouty family, which I borrow from Sir Alfred Garrod, will at once dispose of the foolish idea that gout seldom proves fatal, and then only in extreme old age. He describes his patient thus:—" A gentleman forty-eight years of

age, whose health has been good with the exception
of attacks of gout, which commenced at the age of
thirty-six in one great toe. The attacks gradually
became more frequent and more prolonged, so that
he was scarcely ever free from them." Of this man's
family he says:—" The father had very severe gout;
the mother, when seventy years of age, began to
suffer from it; he has had six brothers, of whom one
died of very severe gout, and was crippled from chalk
deposits in both upper and lower extremities; another
had severe gout and chalk-stones, and died of albumi-
nuria; a third had gout and paralysis, of which he
died; a fourth had gout, and died of erysipelas;
a fifth died of gout, complicated with some urinary
affection; and a sixth is alive, but suffers from gout
in the same way as the patient himself." * Here Sir
Alfred's patient was only forty-eight years of age, yet
he had suffered for twelve years, and was "scarcely
ever free from attacks," while of his brothers five were
dead of gout, and the only one living was crippled
like himself.

In those of the gouty diathesis, death often occurs
early in life from an attack of some of the acute
inflammatory disorders, to all of which the gouty are
specially susceptible. Later in life death is common
from organic disease of the kidneys, from heart disease,
asthma, and apoplexy, the rottenness of the blood-
vessels in the brain rendering the gouty more liable
to this last-named affection than any other class.

* *Op cit.*

Besides causing apoplexy and paralysis, this diseased condition of the blood-vessels in the brain often causes mental disorder, not seldom terminating in complete dementia.

Although gout seldom attacks the female, it is frequently transmitted through the female to the males of the next generation; consequently it is of great importance that the man who has inherited a predisposition to gout should not marry the daughter of a gouty family, for in doing so he makes it doubly certain that the children shall inherit the disease tendency, and that in an aggravated form. Of course all men should avoid alliance with the scrofulous and the rheumatic, but with those of the gouty diathesis such an alliance is specially dangerous to the offspring, who will probably develop painful and deforming diseases of the bony framework of the body.

Again, the man who has inherited a tendency to gout, besides marrying a healthy woman, should, if he marry at all, marry young, for he thereby, as Sir Spencer Wells has shown, vastly reduces the chances of his children inheriting the disease tendency in all its strength. Indeed, if the gouty were to unite only with the untainted, and children were only begotten prior to the outbreak of the disease in the parent, it is probable that, with attention to dieting and exercise, the disease tendency might eventually be eradicated, even in families where it has " run ' for many generations.

CHAPTER XVI.

EXCEPTING gout, there is perhaps no other disease the hereditary character of which is more generally recognised by the multitude than rheumatism. This disease, like gout, appears to depend upon the existence in the system of certain poisonous matters, but exactly how this poison is produced, or why it should be retained in the system in one individual and not in another, medical science is not yet able to explain. It is suspected that some interference with the natural excretive action of the skin, probably having a nervous origin, may have much to do with the presence of the poisonous material within the system; but as to this there is no absolute proof. It has been noticed for ages that in some persons checking the cutaneous action by severe or repeated chills favours the appearance of this disease, and also that indulgence in certain foods and drinks acts in a similar manner. Some believe that these—interference with the action of the skin and indulgence in improper food—persisted in, aid largely in building up in the healthy the constitutional condition necessary for the development of the disease, but beyond the supposition we can hardly at present go. Why a chill, which to one person is harmless, should in another be followed by high fever, swelling of the joints, excruciating pain, and later by disease of the heart, we cannot explain. All we can say is, that a special temperament predisposes the one to this suffering and sickness, and the other escapes because he is not possessed of that peculiar temperament.

The peculiar temperament here referred to is called the rheumatic diathesis, and it is hereditary. Authorities agree that in about a third (30 per cent.) of all cases of rheumatism, hereditary predisposition can be traced, but that these figures represent anything like the real amount of hereditary taint existing among such cases, I do not for a moment believe. Of course the rheumatic diathesis, like every other pathological character, has been and is being acquired by certain individuals, and some of the cases of rheumatism which turn up to-day may be of those in which the necessary abnormal temperament has been acquired within the lifetime of the individual. Still I cannot believe that the acquisition of so grave and far-reaching a diathesis as the rheumatic within one lifetime is common, and that it occurs in anything like 70 per cent. of all the cases of rheumatism at present met with, I deny, and shall prove to be untrue.

When observers tell us they can trace heredity in 30 per cent. of all the cases of rheumatism they meet with, what do they mean? Simply this, that in that proportion of cases they have discovered that a parent or other near ancestor has actually suffered from rheumatism.   Now this is satisfactory only if we recognise the rheumatic diathesis as an acquired diseased condition, transmissible from parent to child, but not transmutable.   But if we recognise this rheumatic diathesis as a degenerate condition, affecting the whole economy, and therefore transmutable, which *may* appear in the next and following generations unchanged, but which may in future generations be transmuted to gout, epilepsy, scrofula, or insanity, then I say such estimate of its hereditary character is misleading.   Let us treat it as other hereditary degenerations.   Is the child of the insane parent who is an idiot, an epileptic, or an instinctive drunkard not to be recognised as the inheritor of the parental infirmity?   Are the children of the confirmed epileptic, who are respectively idiotic, deaf-mute, drunken, and insane, to be considered free from the family taint because their degeneracy has not proclaimed itself in them by the identical symptoms found in the parent? Assuredly not.   Why then should we refuse to recognise hereditary taint in the victim of the rheumatic diathesis whose parents have shown gout or scrofula, apoplexy or insanity, or some other of the degenerate conditions by which decay of the stock makes itself known ?

In the family history of a patient of my own, which I have given at page 186 in the chapter on cancer, we have seen how such apparently diverse symptoms

of family degeneration as infantile convulsions, suicide,
epilepsy, cancer, consumption, and insanity are allied,
and often spring from a common stem. In the following family history—also that of a patient of my own
—we shall see that rheumatism is no exception to the
rule that all family degenerations are transmutable,
the form the disease assumes in the individual only
proclaiming the organs or tissues which the family
blight has specially attacked in each particular case.
Of course the tendency, more or less strong in all
cases, is to reproduce in the offspring the particular
blemish of the parent, and we have seen that in some
cases, as in the suicidal impulse, the identical abnormality is often transmitted ; but to admit hereditary
influence only in such cases as this occurs, is to miss
the whole lesson taught by heredity.

J. G. A.'s FAMILY HISTORY.

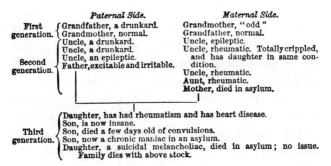

First generation. { Grandfather, a drunkard. Grandmother, normal. — Paternal Side.
Grandmother, "odd" Grandfather, normal. — Maternal Side.

In the above stock we have on the paternal side
the drunkenness of the grandfather transmitted un-

changed to two of his sons, and in another transmuted
to epilepsy—a very common change—while in the
fourth the family blight was only represented by
irritability.   On the maternal side, the " oddity "
of the grandmother, a purely neurotic character,
becomes deepened to epilepsy in a son and insanity
in a daughter, while in the three remaining members
of the family the character of the degeneration is
transmuted from the neurotic to the rheumatic diathesis, one son being totally crippled by the rheumatic affection, and having a daughter already crippled
from the same cause.   It is also to be noticed in this
family that the two neurotic children never suffered
from rheumatism, and the three inheriting the rheumatic diathesis did not show any symptoms of the
neurotic.   The union of members (both neurotic)
of these degenerate families produced five children,
three sons and two daughters, not one of which
escaped the family taint.   One escaped the insane
or neurotic diathesis only to inherit rheumatism of
a virulent type ; infantile convulsions, suicidal melancholia and chronic-mania branding the others as the
offspring of a decaying stock.

Some may think from rheumatism to insanity
rather a long cry, and look upon the appearance
of both in the above family as a mere coincidence ;
but a very brief examination of the facts will conclusively prove the near relationship existing between
these two apparently distinct symptoms or forms of
constitutional decay.

We would first notice that it has been known for centuries that the rheumatic as well as the gouty were themselves specially liable to mental disorder, the older writers correctly pointing out that the mental aberration most frequently took the form of melancholia, with more or less stupor in the rheumatic, and acute mania in the gouty, and that in both cases it at times terminated in hopeless dementia. This being so, we are not surprised to find " Rheumatic Insanity" and " Rheumatic and Gouty Insanity" appearing in various classifications as recognised varieties of mental disease. Not only are the rheumatic and gouty specially prone to inflammations of the membranes of the brain, often causing mental disorder which may become permanent, and later in life to paralysis and dementia following apoplexies, and the deep and hopeless melancholia found associated with diseased blood-vessels in the brain ; but they are also liable above others to the ordinary forms of mental disease. Dr. Clouston of Edinburgh has published * some interesting cases of this rheumatic insanity, and has strongly insisted upon the rheumatic origin of the mental disorder.

All this, however, only relates to those who have or have had rheumatism, and it still remains for me to show that not these only, but their relatives also, are specially liable to nervous disease having mental symptoms. I would trace the relationship between the rheumatic and the neurotic diatheses thus : —

* *Journal of Mental Science,* July 1870.

16

Between rheumatism and chorea, or St. Vitus's Dance, a disease which attacks great numbers of children between the ages of seven and fifteen, though not by any means confined to that period of life, there is a most remarkable connection. Children who have suffered from rheumatic fever, or whose parents are rheumatic, are eminently prone to this disease. M. Sée found that 56 per cent. of all the cases of rheumatism admitted into the Hôpital des Enfans were complicated with chorea, and the late Dr. Hillier stated that in 60 per cent. of his cases of chorea, either the patients themselves, or one of their parents, had been rheumatic. Dr. Copland first pointed out this remarkable connection between rheumatism and chorea, and many attempts have since been made to explain it. Drs. Kirkes and Hughlings Jackson have advanced a theory which is applicable in some cases, but no one has yet advanced a theory which is applicable to all.

Here, then, we have positive evidence of an intimate relationship existing between rheumatism and chorea. Now let us see what chorea is. Chorea is a purely nervous disease, whose symptoms are convulsive muscular movements, hysterical mental disorder, and, in chronic cases, permanent impairment of the intellect. It generally comes on suddenly, and when any cause is given by the child's friends, it is usually "fright," which simply means that the child is of a nervous temperament. One writer on chorea says:—"It is admitted that in a large proportion of cases there is a neuropathic state which antedates and predisposes to chorea. This state is often manifested in the family history by proneness to affections of the nervous system, and in the individual by a highly excitable state of the emotions, so that he evinces joy, grief, or anger from slight causes. All writers admit that there is often an inherited predisposition to chorea." * In fact, chorea may be said to stand midway between the rheumatic and the neurotic diatheses. We have seen how closely related it is with the former, and if we inquire, we shall discover that it is equally nearly allied on the other side with the later.

Dr. Radcliffe made careful inquiry into the relationship existing between chorea and nervous disease generally, and found that in 56.2 per cent. of all his cases of chorea, "the father, mother, brother, or sister had been, or was the subject of one or other of the following disorders :—paralysis, epilepsy, apoplexy, hysteria, or insanity." This is how the case stands, then : We first discover that in about 60 per cent. of all cases of chorea, the patients themselves, or their parents, are rheumatic. Next we discover that about the same percentage of all patients suffering from chorea have had a father, mother, brother or sister who has shown unmistakable signs of the neurotic diathesis—in other words, has actually suffered from paralysis, epilepsy, apoplexy, hysteria, or insanity. And what does all this prove ? Simply

* J. Lewis Smith's "Diseases of Infancy and Childhood."

that the rheumatic and neurotic diatheses, which at first sight appear distinct and far removed from each other, are in reality very closely allied—that they are, in fact, interchangeable.

Much further evidence might be advanced in support of this relationship between the rheumatic and neurotic diatheses, but I will only trouble the reader with the following:—When rheumatic disease attacks the joints with severity, we call it rheumatic fever or acute rheumatism; when it attacks the muscles of the back, we call it lumbago; when it attacks the great nerve of the leg, we call it sciatica; and when it attacks the other smaller nerves, we call it neuralgia. Now the connection between severe neuralgia and insanity has been pointed out by scores of writers, and neuralgia is looked upon by the whole medical profession as an unmistakable sign of the neurotic temperament, yet it is nothing more nor less than a certain form of rheumatism. Dr. Maudsley comments on the relationship existing between insanity, chorea, and neuralgia thus:—"Neuralgia in the parent may manifest itself in the offspring in the form of a tendency to insanity, and every experienced physician knows that if he meets in practice with a case of violent neuralgia . . . he may predicate the existence of insanity in the family, with almost as great confidence as if the patient were actually insane. How it is we know not, but so it is that a certain form of neuralgia owes its origin mainly to a neurotic inheritance. Chorea, again, which has been described fancifully as 'an insanity of the muscles,' is a nervous disease which exhibits sometimes a close relation of descent to insanity or epilepsy; and in children descended from families in which there has been much insanity we meet occasionally with diseased phenomena that seem to be hybrids between chorea and epilepsy, or between chorea and insanity, and which pass finally into one of these more definite ruts of convulsive action." *

I think I have now proved sufficiently clearly the relationship between rheumatic and nervous disease.

I need hardly say the transmutability of all other constitutional degenerations hereditarily transmitted could be equally clearly demonstrated, but I shall not here essay the task. My only excuse for interpolating the last few tedious pages is the desire to point out how nearly allied are so apparently widely separated degenerate conditions as rheumatism and insanity, and to rid the mind of the reader of any lurking suspicion that the family histories given at pages 49 and 231, may be merely those of specially unfortunate families, in which the appearance of several diseases were mere coincidences, instead of the varying signs of an all-pervading decay in the family stock.

Rheumatism is in itself a most severe and painful affection, and although in a large proportion of cases not directly fatal, it is responsible for a very large number of deaths from heart disease, kidney disease,

* "Responsibility in Mental Disease."

asthma, apoplexy, syncope, dropsy, and various affections of the lungs, all of which depend upon and are secondary to disease of the heart. The great danger in all cases of acute rheumatism, and indeed in sub-acute cases too, is the development of heart disease. This occurs in about 50 per cent. of all cases, and is generally the foundation of a condition which ultimately, directly or indirectly, destroys the life of the individual.

I may here remark that there is a form of heart disease which " runs in families," and which frequently appears without the person ever having suffered from either acute or sub-acute rheumatic attacks. Nearly all such cases I attribute to the action of the rheumatic poison upon the heart. If the histories of the families in which this form of heart disease occurs be inquired into, it will be found that rheumatism, intractable neuralgia, or other sign of family decay, such as consumption or insanity, is common in the stock. Here is the history of such a family, every member of which was personally known to me :—Father has suffered for many years from most intractable neuralgia of the head and neck ; no heart disease. Mother became insane at about sixty years of age, as did her father ; no heart disease. There were five children as follows :—1. Son, has grave heart disease ; never suffered from rheumatism ; is so eccentric that he cannot earn his living. 2. Daughter, has grave disease of the heart, and never suffered from rheumatism. 3. Son, has been on the verge of insanity,

4. Son, died of consumption at thirty years of age; and 5. Daughter, eccentric; has heart disease; never had rheumatism.

Rheumatism is so nearly allied to gout that some observers have insisted upon their being identical. Mr. Hutchinson has said:—" Gout is chronic rheumatism made special. . . . Gout is probably chronic rheumatism *plus* a dietetic derangement. Arguments in favour of this view are found in tracing the family history in cases. In most instances of gout, the family history will show chronic rheumatism in some members, frequently on the female side, the males being liable to the fully developed gout, with chalky deposits." * It is certainly true that these two affections do often thus appear in different members of the same family, yet the diseases are not any more identical than are cancer and insanity, which, as we have seen, also frequently appear in different members of the same family. The appearance of rheumatism and gout in members of the same family is to be explained exactly as is the appearance of cancer and insanity, viz., by the transmutability of diatheses. Rheumatism must not be looked upon as a vulgarised form of aristocratic gout; it attacks all classes with charming impartiality, and while it is to be found among those who know the meaning of hunger and hardship, it is to be found equally well developed among those who have for generations been fondled in the lap of luxury.

* *British Medical Journal*, June 2, 1877.

Those of the rheumatic diathesis are very prone
to internal inflammations—as of the sac which holds
the heart, the membranes of the brain, the pleuræ,
and later in life to disease of the kidneys, and also
of the bones, ligaments, and joints, which frequently
terminates in complete crippling and great deformity.
Disease of the great blood-vessels is also very common, and not a few fatal cases of aneurysm and
other like diseases have their origin in the rheumatic
state.

In the vast majority of cases rheumatism makes
its first appearance between fifteen and thirty years
of age, but it may occur in early life, and although
it is comparatively rare before five, it occasionally
occurs, leaving heart disease behind, during the second,
or even the first, year of life.

As to marriage, it is clear that, other things being
equal, the person having the rheumatic diathesis is
not as fit and proper a candidate for matrimony as
the person who is free from this taint of unfitness. We have seen how it may be transmuted, in
transmission to the children, to chorea, neuralgia,
paralysis, heart disease, insanity, or other of the
symptoms of family decay. It is therefore advisable
that those who wish to live in posterity should avoid,
so far as possible, intermarriage with those who
themselves, or whose immediate relatives, have this
diathesis well marked. And while this duty devolves
upon the healthy, who desire their children to escape
avoidable suffering, as far as is possible, it comes

with a force a hundred times increased upon those who have inherited the same or any other degenerative diathesis. Let me say again, that the marriage of persons who have both inherited the same disease tendency vastly increases the chances of the children inheriting the same tendency, and that in an aggravated form; while it is equally certain that the intermarriage of persons who have inherited different disease tendencies, is hardly less dangerous to the offspring. Indeed I question whether this last is not the more dangerous, for while in the former it is common for some of the children to escape the family blight, it is rare indeed in marriages of the latter description for any of them to escape one or other of the parental tendencies to disease, or some combination of both, more terrible than either.

Dr. Benjamin W. Richardson says :—" The worst intermarriages of disease are those in which both parents are the inheritors of the same disease. . . . Intermarriages of distinct diseases are hardly less dangerous. The intermarriage of cancer and consumption is a combination specially fraught with danger. . . . The intermarriage of rheumatic with consumptive disease is productive of intermediate maladies, in which the bony framework of the body is readily implicated. Children suffering from hipjoint disease—*morbus coxarius*—are common examples of this combination. Hydrocephalic children are frequent results of the same combination." *

* "Diseases of Modern Life."

Of course Dr. Richardson means diathesis, or disease tendency, when he here speaks of the inter-marriage of *disease*. The examples he gives of the result of the union of the rheumatic and consumptive are what we might expect; the rheumatic condition having more or less devitalised the bones, joints, and connective tissues generally, these structures are naturally seized upon by the scrofulous disease germs as the most vulnerable point of attack.

CHAPTER XVII.

EARLY MARRIAGES : THEIR EFFECT UPON THE CHILDREN.

"The young man who marries before his beard is fully grown, breaks a law of nature and sins against posterity."—T. S. CLOUSTON, M.D.

THIS is a matter of grave importance alike to the moralist, the economist, and the physiologist. I do not purpose here touching upon it from a social or economic point of view. That aspect of the question I leave for the consideration of the social reformer, who will have difficulty in discovering a field in which his energies may be more profitably expended. Few will deserve better of their country than he who succeeds in staying to any appreciable extent the reckless rush of precocious, ill-developed children into matrimony which is at present going on among our people.

This question of early marriages was brought before the London Diocesan Conference in 1889, and it was then agreed on all hands that the evil had grown to such an extent as to render some reform in the marriage laws urgently necessary. Before and since that time the matter has frequently been under consideration by various organisations, but up to the present nothing has been done to alter the law which permits boys and girls, however sickly and ill-developed, undertaking the most important duty which falls upon the citizen, viz., the renewing of the population.

The marriage contract is by far the most important transaction which the ordinary citizen enters into during the course of his or her natural life—important alike to the individual and to the state; yet according to the law as it at present stands, the minor, who is not incapacitated by idiocy or raving madness, can at almost any age become a party to a contract of this nature which shall be binding during the remainder of his or her natural life. That this is a grave mistake I think all who seriously consider the subject must agree. In ordinary contracts, unless the thing contracted for can be proved to be "a necessary," the minor can in almost every case successfully plead his minority at the date of contract as voiding the engagement. Yet in the case of the marriage contract, where the thing contracted for can never by any possibility be a "necessary," and where the lifelong happiness of two persons and the well-being of a possible family are involved, the under-age contractor can get no relief, however designing and untruthful the other party to the contract may have been.

In this question of child-marriage the moralist and the economist both consider themselves injured parties, but neither the one nor the other has a tithe of the solid ground for complaint that the physiologist has.

Arguments, plausible if unsound, might be advanced tending to show that early marriages make for morality, and something might also be said in answer to the objections of the economist; but there is not a single word to be spoken in mitigation of the sweeping condemnation which the physiologist is compelled to pronounce upon all marriages of the immature and the senile.

To the superficial observer it may appear that every marriage must enrich the state, and that early marriages must lessen the amount of sexual immorality, but inquiry will prove conclusively how fallacious are those views.

Early marriages certainly tend to the production of large families, but then a family, to be a source of wealth to the state, must at least be self-supporting, which is exactly what the feeble, degenerate children of the great mass of our early marriages are not. They are brought forth ill-developed and unhealthy; their immature, improvident parents are unable to either feed or educate them as they ought to be fed and educated; hence, instead of being a source of wealth to the state, they prove a serious drain upon her resources. A large percentage of these miserable children succumb during infancy, but a great number drag out a pitiful existence, only to become inmates of our workhouses and infirmaries, our asylums and prisons, and, after being supported at the public expense for a longer or shorter period, to die prematurely, leaving the state poorer than they found it and no better. It is indeed a small percentage of the children of the immature that ever become robust, useful, self-supporting citizens.

Again, can any one who looks beyond the immediate present seriously argue that morality can possibly be the gainer from such marriages as those under review ? Who will venture to say that the immorality which might possibly be indulged in by the individual during his minority because of his unmarried state can approach that which must be the natural outcome of the presence in society of his half-dozen ill-developed, half-educated, half-starved children ? Where do immorality and vice assume their most hideous forms ? Is it not in the dens where the wretched children of these immature, improvident, and impoverished parents are huddled together in the slums of our great centres of population ? But now I am encroaching upon the domain of the sociologist, a thing I promised not to do, and an aspect of the question with which I am not competent to deal. Let us at once, then, and briefly, consider the matter from the standpoint of the physiologist, and learn how these early marriages affect the standard of health and the vitality of the community.

It is impossible that individuals, male or female, who have not themselves reached maturity, can beget or bring forth a fully developed, healthy offspring. The child has only that quantum of vitality which has been conferred upon it by its parents, and should they be deficient in vital power, of necessity so also must be the child. The deficiency of vital energy in the

parent or parents may arise from various causes. It
may have been lost from exhausting disease, from
vicious excesses, from approaching senility, or it may
never have been had to lose, as is the case in the
immature; but whether never possessed, or from whatever cause lost, its absence in the parent is equally
serious to the child. That which the parent has not,
he cannot entail upon his children. The children of
the immature lad, the enfeebled invalid, and the
worn-out roué are of necessity born into the same low
grade of vital poverty.

From very early times it has been noted, of man
and animals alike, that the young begotten or brought
forth by the immature has been wanting to a greater
or less extent in strength, stamina, and courage—in
general development, in fact. No breeder of stock
would permit his mares, heifers, or ewes, however
healthy, to bring forth young before they had arrived
at maturity, nor would he permit an immature male
to impregnate the females of his herds or flocks.
When such does occur, the offspring is invariably
small, weedy, and not worth the trouble its bringing
up entails. "The young of' animals not yet fully
developed are small and stunted, incapable of perfection: it is observed in foals, lambs, goats, calves, &c.,
born of very young parents; they remain weak, lymphatic, and functionally inert." * Here is a case in
point which recently came under my own observation.
A young sow was impregnated by a mature boar, and

* "A Physician's Problems," by Charles Elam, M.D.

when eleven months old brought forth seven pigs. They were tiny, ill-developed things; some died immediately after birth, others within a few days, and all of them succumbed within five weeks of the time of birth. The strength of the mother was considerably taxed by the operation, but she recovered, and having gained maturity, she had well-developed and healthy offspring by the same father. It is a common practice in many districts to destroy the first litter of puppies brought forth by a bitch, and if you ask the "fancier" why he does so, he will inform you that "the first litter are never any good; they are sickly, and seldom get through the distemper; they have no pluck, and they never come to anything."

In the human family the same rule holds good. The children born of parents who have not themselves reached maturity are markedly inferior to those born under like circumstances of mature parents. Aristotle remarked that in those cities of Greece where it was the custom for young people to marry before maturity, the children were puny and of small stature. Montesquieu observed the same fact when, in France, the fear of conscription induced great numbers of young people to marry long before the proper period: the unions were fruitful, but the children were small, wretched, and unhealthy. According to M. Lucas, the same occurred in 1812 and 1813.

A vast number of these children of the immature are born prematurely, to the great danger of the imperfectly developed mothers; a much larger percentage of them are idiotic, dumb, blind, scrofulous, and otherwise imperfect and deformed than the children of parents generally; they have a less firm hold on life than the children of mature parentage, and many succumb to scrofulous and nervous affections, a great number dying of convulsions before or during the period of the first dentition. As a class, such children are decidedly not long-lived, and those who do attain the age of maturity are generally delicate and under-sized physically, often obtuse, and more or less dwarfed mentally, if not distorted at least blunted morally, and are wanting in spirit, energy, and courage.

This last mentioned trait—lack of courage—is peculiarly characteristic of the offspring of immature parents, as well in the human family as in the brute creation. For instance, dogs the offspring of immature parents, whatever the ferocity of the breed, are timid, fearful creatures, that can be taught nothing; spiritless curs, that are absolutely worthless, and are as a consequence but rarely permitted to long survive their birth. In the human family this want of courage has always been recognised as a failing of children of immature parentage. To be called the " son of a boy " or the " child of a girl " is synonymous with being called a coward. Thus Shakespeare, who knew something of everything, makes Macbeth exclaim when addressing Banquo's ghost :—

> " What man dare, I dare.
> Approach thou like the rugged Russian bear,

17

> The armed rhinoceros, or the Hyrcan tiger,
> Take any shape but that, and my firm nerves
> Shall never tremble : or, be alive again,
> And dare me to the desert with thy sword ;
> If trembling I inhibit thee, protest me
> The baby of a girl." *

This absence of courage and virility in the children
of the immature is well illustrated in the result of
Marro's investigations.   He found that an astonish-
ingly large percentage of the insane and thieves were
the children of immature fathers, while the same
class was but poorly represented among murderers
and sexual offenders, where courage or ferocity and
animal vigour were necessary (*vide* page 262).

Authorities are unanimous in agreeing that the
children of mothers under twenty, and of fathers
under twenty-four, are, as a class, less robust mentally,
morally, and physically, than children of parents in
their prime.   M. Joseph Korosi, of the Buda-Pesth
Statistical Bureau, has made more extensive inquiries
upon this subject than any other investigator, and
his conclusions agree closely with those of Marro
and others, viz., that immature parents bring forth
a degenerate stock, in which the percentage of
idiots, cripples, insane, consumptives, criminals, &c.,
is immensely larger than in the children of mature
parents.   The most perfect, robust, and long-lived
children are those of fathers between twenty-five and
forty, and of mothers between twenty and thirty years.

The late Dr. J. Matthews Duncan, who enumerated

* "Macbeth," Act III. Scene iv.

among the evils arising from premature marriages "abortions, early death of children, excessive families, twins, triplets, idiots, and sterility," says :—" In woman the age of maturity is twenty to twenty-five ; in men it is later, probably by at least five years ; and you will pardon the interpolation here of the reflection, well worthy of being fully dwelt on, that this late ascertained physiological law tallies with the old and wisest counsels as to the nubility of men and women—a part of the grand subject of morals and medicine. At this age a woman has the lowest risk of sterility, the greatest likelihood of having healthy children that will long survive, the greatest likelihood of herself surviving childbirth, the lowest risk of having abortions, of having excessive family, of having plural pregnancy, and of bringing forth idiots." *

Even the children of healthy and well-fed boys and girls of the better classes are decidedly inferior to the children of mature parents. That has been proven. What, then, must we expect from the premature marriages of the stunted, pale-faced boys and girls of our great cities ? In them the deteriorating influences of city life have produced a class largely lacking in vitality—a class which is, as we have seen, actually in process of decay, and which, even at its best, could not produce an offspring with the normal quantum of vital energy. Yet this is the very class among which we find premature

* *Lancet*, March 23, 1889.

marriages most common. The pale, wan, sad-eyed
factory-hand, or other city toiler of either sex, without a single penny put past for the rainy day, at
any age from twelve and fourteen upwards, undertakes all the cares and responsibilities of the married
state, with a lightness of heart which can only arise
from ignorance or carelessness, or both. To these
children, children are born, two-thirds of whom,
happily, die in infancy, while the other third live
a charge upon their fellows. No doubt can exist
that these child marriages take almost equal rank
with want of fresh air and sunlight, poverty and
drunkenness, as agents in the production of that
general deterioration and decay, which exterminate
our poor city dwellers within three or four generations.

According to the last Annual Report of the Registrar-
General, there were during the year 1889, 94,040
men of twenty-one and under, married. Of these,
80,905 had gained their majority, 8769 were twenty
years of age, 3576 were nineteen, 728 were lads of
eighteen, 59 and 3 were boys of seventeen and sixteen respectively. On turning to the women, we find
that no less than 42,170 were married of twenty
years of age and under. Of these, only 19,223 were
twenty, 14,129 were nineteen, 7159 were eighteen,
1472 were seventeen, 167 were sixteen, and 20
were fifteen.* But this is not the worst. Had

* In 1888 two brides were fourteen, and one had reached the
wifely age of thirteen years.

these immature persons married robust individuals of mature age, the result would have been much less injurious to posterity than it will be, for the union of an immature with a mature parent is decidedly less injurious to the offspring than the union of the immature with the immature. Upon analysis, however, we discover that 33,526 lads of twenty-one to sixteen years of age married wives of twenty to fifteen years old, and that 33,526 girls of twenty to fifteen years of age took for husbands lads of twenty-one to sixteen years old. Now if we remember that the vast majority of these boy and girl husbands and wives belonged to the working-classes, and that the greater number of them sprang from the ill-developed, prematurely decrepit city dwellers we have been speaking of, we shall get some idea of what suffering and poverty these premature unions caused, what child-death they represented, and what a source of contamination they must prove to the health of the people.

But why, it will be asked, should these thoughtless, improvident boys and girls be permitted to marry? According to the recent Criminal Law Amendment Act, no girl under sixteen can consent to illicit sexual indulgence; her consent in no way mitigates the offence of the man who knows her carnally. Now, if her consent to illicit intercourse be worthless, of what value can be her consent to marriage? It may be argued that in the case of marriage the child is guided by the wise counsels of parents or guardians, but we know as a matter of fact that the advice of parents is not by any means invariably wise when the marriage of their children, old or young, is in question, and also that most of the marriages entered into by girls under sixteen are contracted against the advice and wishes of, and too often without the knowledge of, the parents of the girl. Assuredly if a girl of sixteen be unable to choose upon whom she shall bestow her favours temporarily, she should be deemed unable to choose a partner for life, or to delegate to another so momentous a duty.

Now that the ecclesiastical element forms no necessary part of the wedding ceremony, and the marriage bond has become a mere civil contract in the eye of the law, it is difficult to see why a girl of twelve, thirteen, or fourteen years of age should be able to enter into a life-long contract of a sexual nature, and at the same time be unable to enter into a temporary contract of a similar kind. If the law denies her the right in the one case, it should deny her the right in the other. If she be unable to choose a lover, as she most certainly is, she is unable to choose a husband, and no hardship could possibly arise from forbidding her the right. At present the ages at which the marriage contract can be legally entered into are (1 Jac., c. 1) twelve and fourteen for girls and boys respectively, which is absurd considering how imperfectly developed children are at those ages in these countries. I do not suppose any one would seriously object if it were enacted that girls under eighteen and lads under twenty could not legally make a binding marriage contract. If such a law were established and could be effectually carried out, it would raise the standard of morality among the lower ranks of society, reduce appreciably the grinding poverty and overcrowding in the industrial centres, and materially increase the average vital capacity of our people.

CHAPTER XVIII.

LATE MARRIAGES : THEIR EFFECT UPON THE CHILDREN.

"Filii ex senibus nati, raro sunt firmi temperamenti."
—Scoltzius.

Dr. Marro fixes the period of decadence in fathers at from forty years onward. M. Korosi, after inquiry in 24,000 cases, says above forty, fathers tend to beget weak children. The most healthy children have mothers below the age of thirty-five. During the year 1889 no fewer than 11,516 men of forty-five to upwards of eighty-five years of age were married, and 11,148 women of forty to eighty years became wives in England and Wales. Of these wives 3,472 were fifty or more years of age, and from these posterity will suffer small injury. But the marriage of the remaining 7,676 women of between forty and fifty, and 11,516 men already some way on the downward path to decay, must have a great and gravely injurious effect upon the generation in which their offspring shall appear.

Just as the immature parent who has not reached perfect development produces a degenerate offspring, so the elderly parent, the tide of whose vitality is on the ebb, brings forth children more or less

imperfect in mind and body. And the further the
vital tide has receded in the parent, the farther
removed from the high-water mark of perfection
will be the children.

I have said, in the preceding chapter, that whether
the lack of vitality in the parent or parents arises
from immaturity, disease, exhaustion, or old age,
the effect upon the offspring is the same. This is
true, yet it is not the whole truth. The late Dr.
Griesinger of Berlin, speaking of idiocy, says:—" In
families where epilepsy, mental disease, paralytic
affections, deaf-dumbness are frequent, idiocy is also
observed to be common. Frequently it occurs as a
mere partial phenomenon, as an individual manifes-
tation of a general degeneration of the race: thus
we see, in a number of brothers and sisters, one or
two idiots, together with others who are small, in-
completely developed, ugly, and sterile. These degenerations are observed in families . . . where the
parents have been too old or too young." * Here
the children of the immature and the senile are
classed together, as is usual. They have much in
common. Both are sadly wanting in vital power
and in perfection of development, physical, mental,
and moral. Hence we have a vastly greater percentage of idiots, deaf-mutes, insane, epileptics, thieves,
cripples, dwarfs, drunkards, and sterile individuals
among both classes than we can discover among the
children of mature parentage. But although they

* "Mental Pathology and Therapeutics," Syd. Soc.

have all these and other imperfections in common, there are many points of difference which, if not of very great practical importance, are deeply interesting. Let us briefly consider some of these.

In the first place, although early death is much more common among both these classes than among ordinary children, there is not the same terrible mortality during the first year of life among the children of the senile that occurs in those of the immature. A much larger proportion of the former live to years of maturity, although they seldom or never reach old age. Indeed, they may be said to be old from their birth, for they have many of the characteristics of the aged while still children. Prosper Lucas * and other writers have given some remarkable descriptions of the aged aspect of such children.

Another most interesting and curious difference is, that whereas both are specially prone to mental imperfection, the insanity or mental distortion does not take the same form with equal frequency in the two classes. Thus among the offspring of the immature, we find blank idiocy very much more common than among that of senile parentage. Among the latter, congenital mental defect is very common also, but it does not nearly so often take the form of utter vacuity as it does in the former. In the one case you generally find blank idiocy, accompanied, it may be, with deaf-mutism, blindness, epilepsy, or

* " Traité philosophique et physiologique de l'Hérédité naturell."

some such bodily imperfection as squint, paralysis, cleft-palate, &c., while in the other, the young of the senile, you do not meet with the same complete absence of mental power, nor the same amount of such deformities as club-foot, cleft-palate, paralysis, squint, blindness, &c.

The distinctive characteristics of the two classes might be roughly summed up as follows:—The children of immature parentage are specially liable to death during infancy from wasting, scrofulous, and convulsive affections. They are liable in a remarkable degree to idiocy and imbecility of a low type, and to physical deformities and imperfections. Large numbers of them succumb to tubercular disease about the ages of puberty and adolescence, and few of them attain even advanced middle age. The genital organs are ill-developed and often deformed, and a great number of them are sterile. They are also notorious for their lack of energy and courage. Hence the class of criminals to which they give the greatest number of recruits is that of thieves and other petty offenders.

The children of the senile are as a class ugly, small of stature and stooping, which, together with the absence of subcutaneous fat, gives them a look of old age while still young. Idiocy is less common amongst them than weak-mindedness amounting to imbecility, which is often accompanied with more or less perversion of moral feeling, and a plentiful supply of deep low cunning. Many of them die between the ages of puberty and adolescence of tubercular disease, few of them live past middle age, and great numbers of them ultimately become insane and criminal. They are nervous, irritable, passionate, and horribly cruel, and are the perpetrators of most of those fiendish barbarities, the recital of which from time to time shocks the civilised world.

To put it shortly, the children in each case are like their parents. The immature parent is only partially developed mentally and bodily, and like him his child is wanting both in mind and body. On the other hand, the aged parent, though deficient in physical vigour, is often ripe mentally—for the mental faculties flourish long after the bodily vigour has begun to wane. Consequently, we find in his children mental ability curiously mixed with that peevish irritability so typical of the aged, and distorted as we should expect to find it imprisoned in a prematurely decrepit physical organisation.

This I draw from my own observation and experience after many years' intimate association with large numbers of the idiotic, imbecile, and insane, *and their relatives*, together with some study of those brought before our criminal courts for judgment. Unfortunately I cannot present statistics of my own on the subject, but am convinced that if I could they would support the views here broadly expressed, as do the admirably arranged figures of such eminent investigators as Marro, Langdon Down, and Korosi.

A glance at the accompanying diagram which I

take from Mr. Havelock Ellis's excellent work "The Criminal," * will show how far the inquiries of Dr. Marro go to support the views expressed above.

This plate shows the ages of the fathers at the period of conception of criminals as compared with ordinary persons and with the insane. Marro divided the fathers into three groups, viz.: (1) the immature, to which he considered all under 25 years of age to belong; (2) those of the period of maturity, which he put down as from 26 to 40; and (3) those in the period of decadence, which included all fathers over 40 years of age at the time the child was conceived. In the plate the first column in each group represents the proportion of the children of immature parents, the central column those of mature fathers, and the last the proportion of those of decrepit fathers.

The first thing impressed upon us by this plate, is the fact that "Criminals in general" are midway between ordinary persons and the insane as regards healthy parentage. The percentage of persons of mature parentage among normal persons is 66.1, this is reduced in the criminal class to 56.7, and again in the insane to 47.0; each of these decreases being accompanied by a corresponding increase of those of immature or senile ancestry, or both.

In the next place, we notice that the children of immature fathers are but poorly represented among murderers, sexual offenders, and sharpers, which goes to corroborate my assertion that they are wanting in

* "The Contemporary Science Series." Walter Scott.

Diagram showing Proportion of Immature, Mature, and Senile Parentage in Normal Persons, the Insane, and Criminals.

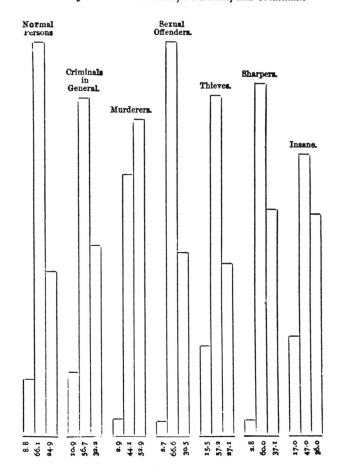

courage, ferocity, animal vigour, and mental power
generally. On the other hand, they are largely represented among thieves, whose occupation demands
the possession of neither courage, energy, nor
brains.

Of those begotten by fathers in the period of
maturity we note that sexual offenders are the only
class which at all approach the normal persons in
respect of mature parentage. We also notice that
amongst murderers, those of mature parentage are
actually fewer than among the insane themselves.

Of those of senile parentage the most notable fact
is the remarkable number of murderers which belong
to this class. Of all the murderers whose ancestry
was inquired into by Dr. Marro, 52.9 per cent.
actually turned out to be the children of fathers who
had passed their prime. This fact corroborates my
estimate of this degenerate class, and fully justifies
the assertion that low cunning, moral perversion, and
heartless cruelty are almost constant characters in the
children of aged parents. This class is also liberally
represented among sharpers and sexual offenders,
where their perverted moral feeling, aided on the one
hand by their cunning, and on the other by their cruelty,
would lead them; but among thieves, where the offspring of the immature is so numerous, they are
found only in small numbers, their percentage being
less than 3 greater than among normal persons. The
cowardice and imbecility of the children of the immature make them thieves and idiots; the cunning and

cruelty of the children of the aged drive them beyond the first shallows of criminality.

From all this it is evident that marriages among the immature and the aged are equally opposed to the maintenance of physical, mental, and moral health among the people. Such unions are the cause of an enormous amount of child suffering and death. From the offspring of these unions are largely recruited the ranks of the idiotic, the epileptic, the insane, the scrofulous, the deaf-mute, the criminal, and every class of the unfit, which are a drag upon the state, a fruitful source of contamination to the health of the populace, and a reproach to our civilisation.

No man having respect for his own health or that of his children should marry until he has at least gained his majority, and the nearer he approaches the age of 25 before he undertakes the responsibilities of the married state, so much the better for both himself and his children.

Men who are past their prime should be very careful in the selection of wives, and should remember that the health and vigour of a *mature* young mother may largely neutralise their own unfitness as fathers. From this last remark it is not to be understood that the aged man should marry an undeveloped girl, as so many do, but a mature woman of 25 to 30 years of age. During the year 1889 no less than 75 men of 45 to 75 married girls of 20 to 15, and no fewer than 193 women of 35 to 50 married men of 21 or less. (These are the figures as given by the parties themselves; and in estimating their value, we must not forget what self-depreciation in this particular the blushing bride of over 35, or the budding Benedict of a decade later is capable of.) The man of 40 to 50 should marry a woman of 25 to 30. From such a union, healthy children may be reasonably looked for. The man of over 50, unless in rare instances where the strength and vigour are exceptionally well maintained, had far better not marry with a view to rearing a family. Rarely indeed will he live in distant posterity.

To advise the female portion of humanity as to the age at which they should abandon all ideas of matrimony, or as to the point at which the youthfulness of a husband becomes objectionable, would, I fear, be a work of supererogation, and for that reason I shall not enter upon it.

18

CHAPTER XIX.

CONSANGUINEOUS MARRIAGES.

THIS is a subject upon which it is necessary to say a few words here, as it is a matter of great importance that something should be known of the effect of the intermarriage of blood relations, more especially by members of families in which exist a tendency to any hereditary disease.

It is popularly believed that the intermarriage of persons nearly related by blood leads to evil consequences in the offspring, and in proof of this it has been pointed out that such imperfections as idiocy, insanity, epilepsy, deaf-mutism, blindness, scrofula, phthisis, paralysis, and various bodily deformities, are much more frequently met with among the children of parents who are close blood-relations than among those of parents who are not so related. Now, that all the imperfections above mentioned and many others are met with among the children of consanguineous marriages is true, and that they occur here more frequently than among the general population is also true, but that this condition of affairs is due to the mere fact of blood-relationship in the parents has been disproved.

It cannot be said that consanguineous unions are repugnant to nature, although custom and religious teaching have developed a repugnance thereto in civilised man. From the early history of mankind we learn that marriages between very close blood relations were both legal and common. " The Persians, Tartars, Scythians, Medes, Phœnicians, Egyptians, and Peruvians not only married their sisters, but their daughters and their mothers. Instances of such marriages among the members of the royal families of antiquity are well known." Later, the laws of the ancient Germans permitted consanguineous marriages of a less glaring kind, as did also the laws of the Arabs until the time of Mahomet; and the Jews, notwithstanding the strict injunctions of Moses, continue them until the present day, as do also that strange people, the Gipsies.

Nevertheless, the frequency of imperfection in the children of such marriages has been noticed from the time of Moses, or earlier, as is proved by the fact that all the great moral codes—Hindu, Mosaic, Roman, Christian, and Mussulman, have alike forbidden such unions. All these laws were evidently founded on the belief, which is still generally accepted by those who have not studied the matter, that the unhappy results so frequently following consanguineous marriages, depended upon the mere fact that the parents were of the same blood. This, however, has

upon inquiry proved to be erroneous; yet this discovery has in no way lessened the practical utility
of the law forbidding the marriage of blood relations.
In fact, there is much more need of a strict observance of this law nowadays among our highly civilised
communities, than there was among the primitive
peoples to whom it was first given.

We have it recorded in Holy Writ that Abraham
was married to his half-sister, Isaac to his first
cousin once removed, and Jacob to his first cousin,
and that the stock from this root flourished exceedingly. Huth, in his "Marriage of Near Kin," cites
several instances of much the same state of affairs
occurring regularly at the present day among certain
isolated communities, such as the inhabitants of St.
Kilda, Pitcairn, and Iceland, without any apparent
evil consequences to the race. It is also certain
that such marriages are, or were, common among the
North American Indians and the South Sea Islanders,
peoples among whom idiocy and other degenerate
hereditary conditions were remarkably rare. In these
cases, however, it must be noticed that we have
peculiarly healthy communities to deal with, and
therein lies the secret of such intermarriage proving innocent of evil to the offspring. Were such
marriages common among the neurotic, decrepit,
scrofulous, and otherwise degenerate dwellers in our
great cities of to-day, the result would be disastrous.

So long ago as 1869 the New York State Medical
Society appointed a committee of its members to

investigate and report upon the influence of consanguineous marriages upon the offspring, and the result of their labours, as published in the *American Journal of Insanity*, 1870, shows clearly that if the family be free from degenerative taint, marriage among its members in no way diminishes the chances of healthy offspring. This conclusion is in perfect agreement with the findings of other and more recent investigators, such as Anstie, George Darwin,* and A. H. Huth. According to these authorities, there is no greater amount of disease or deformity among the offspring of parents related to each other by blood, than among the children of parents not so related, provided the parents be equally free from tendency to disease or degeneration. With a perfectly healthy stock, as every breeder of animals knows, " in and in breeding " may be practised with impunity, but where the stock is tainted with disease or imperfection, safety is only to be found in " crossing."

Where the error lay in the old doctrine, upon which was founded the prohibition of consanguineous unions, was not in asserting that disease and deformity were more often met with in the children of these than in those of other unions, for such is true, but in attributing these unhappy results to the mere fact that the parents were related by blood. Over and above the fact that these consanguineous marriages are almost certain to transmit, in an accentuated form, any defect or tendency to disease already present

* *Journal of Statistical Society*, June 1875.

in the family, there is no physiological reason why such marriages should not take place. Breeders of prize stock frequently breed "in and in," not only with impunity, but with marked benefit. But this fact, while going to prove that it is not the mere blood relationship of the parents which induces the degenerate conditions so often found in the children of consanguineous marriages, can but rarely be advanced as an argument in support of the marriage of blood relations. The stock-raiser only permits the more perfect members of his flocks and herds to continue their kind, and for this reason the "in and in" breeding is innocuous, just as it would be in the human family under like conditions. But where shall we find the perfect human family? At the present time such families are certainly rare. The laws of natural life have been so strained and perverted by our civilisation, that almost every family nowadays has got a taint or twist of some kind, and as all such imperfections are transmitted and rapidly deepened and fixed in the family by the intermarriage of its members, it is best that such unions should in all cases be forbidden.

As we have already seen, recently acquired characters, whether physiological or pathological, are very liable to disappear when the individual bearing such character intermarries with another not having the same character. The natural tendency in all such cases is to revert in the offspring to the normal or healthy type, so that, unless the new character be very deeply impressed upon the parental organism, it is almost certain it will not appear in the offspring, if the other parent have nothing of the character. But when both parents are possessed of the character, whether it be physiological or pathological, this natural tendency to revert to the original is often over-borne, and the character is repeated in an accentuated form in the offspring.

Now, this accentuation of all family characters is what must always happen in the case of consanguineous marriages. If there be any taint in the family, each member of the family will have inherited more or less of it from the common ancestor. Take the case of cousins, the descendants of a common grandparent who was insane, and of an insane stock. Here the cousins are certain to have inherited more or less of the insane diathesis. Even if the taint has been largely diluted in their case by the wise, or more likely fortunate, marriages of their blood-related parents, yet will they have inherited a certain tendency to nervous disease, and if they marry, they must not be surprised if that taint appear in an aggravated form in their children. Some of the children of such parents are generally idiotic, epileptic, dumb, or scrofulous, and the parents marvel whence came the imperfection. It may be, in some cases, that the parents, and possibly the grandparents, of the unfortunate children, have not up till that time displayed any outward evidence of the tendency to disease which they have inherited and handed on to their descendants, and not looking farther back, the parents boldly assert that such a thing as insanity, epilepsy, scrofula, &c., is unknown in their family. They themselves have never been insane—why, then, should their children? In like manner children may be epileptic, blind, mute, scrofulous, cancerous, criminal, or deformed from direct inheritance, and yet the family line be honestly declared to be healthy. Hence the truth of Sir William Aitken's words: " A family history including less than three generations is useless, and may even be misleading."

In consanguineous marriages, then, the danger lies in the strong probability there is of both parents bearing some particular taint of degeneration, which will be deepened in their children, yet which might be escaped if they each married a person not bearing that same, or some allied, character. The blood relationship in itself is innocent. It is the double tendency to disease which brings about the evil to the children. The marriage of two phthisical, or scrofulous, or neurotic persons, whose families know nothing of each other, would be equally pregnant of evil with the marriage of cousins, or even nearer blood relations similarly tainted.

From the foregoing it is evident that the similarity of temperament induced by a common environment, and which I would call " social consanguinity," must be a potent factor in the production of all hereditary degenerations. Living under similar customs, habits, and surroundings, labouring at the same occupation, indulging in the same dissipations, tends to engender like diseases and degenerations irrespective of any blood relationship. Hence it not seldom happens that persons not even distantly related by blood, are, in reality, much more nearly related in temperament than cousins, or even nearer blood relations, who have experienced widely different modes of life. This " social consanguinity " is the great curse which dogs every exclusive tribe and class, and hurries them to extinction. It has largely aided real or family consanguinity in the production of the disease and degeneration which have so heavily fallen upon the aristocracies and royal families of Europe. Therefore the introduction of plebeian blood into the noble family is to be applauded, not only as being poetic, but as being useful. The " lady of low degree " frequently brings with her a heritage of health more valuable than silver or gold. But when she brings gold too, as does the modern American representative of the " Gipsy Countess " of old, there appears nothing left to be desired.

The important part played by this " social consanguinity " in bringing about family degeneration, is well illustrated both positively and negatively in the case of the Jews. This race has permitted the intermarriage of near blood relations from the earliest times up to the present, and such unions have at all times been common amongst them. Yet the Jews have for centuries maintained a physical and intellectual standard quite up to the average of

modern peoples, and that, too, notwithstanding
frequent and cruel oppression. To my mind this
strange immunity from degeneration, after centuries
of consanguineous unions, is only to be accounted for
by the absence of "social consanguinity" among
this people. The Jew has been for centuries a
wanderer upon the face of the earth; a veritable
rolling-stone, though differing from the proverbial
rolling-stone in its one grand characteristic. He is
without a country, therefore without patriotism, and
consequently never a soldier; but, except on the
battlefield, there is no spot on the earth's surface
where wealth and honour are to be won that the Jew
is not to be found—not as a settler, merely as a
temporary sojourner. He has few ties, and he has
no love for one country more than another. So he
moves from town to town from country to country,
from continent to continent in pursuit of wealth,
his wandering bringing about such change of environment that anything approaching the "social
consanguinity" constantly met with among European
aristocracies is impossible.

The same might, to a certain extent, be said of that
other race of wanderers upon whom repeated consanguineous marriages seem to bring no blight—
the Gipsies. In the case of the Gipsies, however,
much of their immunity from degenerative disease
doubtless depends upon their more natural mode of
life, and consequent large store of physical health,
as is undoubtedly the case among savage tribes

and such communities as those of Pitcairn and St. Kilda.

It is certain, then, that consanguineous marriages must be extremely dangerous in communities like our own. Where the laws of natural life are so gravely interfered with they should rarely, indeed, be entered into, if at all. Yet we must admit that just as we can cultivate by "in-and-in-breeding" pathological or degenerate characters such as the insane, gouty, or scrofulous diathesis, so it is possible, by the same means, to foster physiological or healthy characters. The natural law works for good as well as for evil, and it is possible, by means of intermarriage of those belonging to a family noted for some physical or intellectual excellence, to deepen and fix that good character in the family. Thus, the marriage of cousins in whom the literary, artistic, musical, or other talent is prominent will in all probability pro- duce children in whom the particular talent of the parents will be still more strongly marked. A good example of this is found in the numerous family of the Bachs, the musicians, who freely intermarried, and elevated their talent, possessed by all, to the level of genius in some of their members.

In this manner any mental or physical character may be transmitted, deepened with each transmission, and finally fixed as a constant character in the family. Even what might be called accidental characters, or peculiarities, may be seized upon by the breeder and fixed in the family. This is constantly done by breeders of our domestic animals. Of supernumerary fingers and toes—a common deformity in the human species since the days of David, when " Jonathan, the son of Shimeah, the brother of David," killed in battle the Philistine of great stature, who " had on every hand six fingers, and on every foot six toes, four-and-twenty in number ; "—of this deformity Sir William Lawrence says :—" If the six-fingered and six-toed could be matched together, and the breed could be preserved pure by excluding all who had not these additional members, there is no doubt that a permanent race might be formed, constantly possessing this number of fingers and toes." *

This assertion of Sir William might be imagined by some to be extravagant, whereas it is, in reality, well within the bounds of probability. The fact that accidental peculiarities can be reproduced and fixed as constant characters is well known, and constantly taken advantage of by breeders. The best instance of this which I can cite is that of the Ancon sheep. The peculiarities of this variety of the sheep family, which is now largely bred in America, first appeared in a somewhat deformed lamb, born of ordinary parents. " The first ancestor of this breed was a male lamb, produced by a ewe of the common description. This lamb was of singular structure, and his offspring, in many instances, had the same characters with himself : these were, shortness of the

* " Lectures on Physiology, Zoology, and Natural History of Man."

limbs, and greater length of the body in proportion: whence this race of animals has been termed the otter-breed (otherwise the Ancon sheep). The joints also were longer, and the forelegs crooked. It has been found advantageous to propagate this variety because the animal is unable to jump over fences." *

But although consanguineous marriages might thus be used to develop desirable characters in families, such breeding of genius is not to be advocated, for this reason: Few families are physiologically perfect, most have got some unhealthy taint, and while the desired character was being deepened and fixed by successive consanguineous unions, we would in most cases be building up at the same time some pathological character. This latter would increase as surely as the former, and on reaching the necessarily fatal degree would put an end to the family altogether. For this reason, if there were no other, real or family consanguinity should be avoided in marriage, as should also that which I have called "social consanguinity."

From time immemorial it has been known that "the introduction of new blood" has a beneficial effect upon the family or race, and proof of the truth of this old-time doctrine is to be had on every hand, both in the human family and among the brute creation. The most beautiful families of the south are said to be those which proceed from the alliance of Germans or

* "Researches into the Physical History of Mankind," by J. C. Prichard, F.R.S.

Hollanders with the women of the country,* while the
families of Berlin most remarkable for their beauty,
their force, and their intelligence proceed from French
exiles who married ladies of Berlin.† Nearer home, in
Ireland we have positive evidence of the beneficial
effect of " crossing " with fresh blood. In the counties
of Tipperary and Limerick, where great numbers of
Cromwell's English soldiers settled, the people are
noted for their splendid physical development and
wild daring spirit. Again, in Ulster, where the Low-
land Scots planted there by James have blended with
the earlier Celt, the present inhabitants are physically
superior to those of any other part of the kingdom,
while in mental acuteness and energy they are second
to none. The superiority of these mixed races is at
once evident to the traveller in Ireland. On this
point Dr. Prichard remarks :—"In some parts of Ireland
where the Celtic population of that island is nearly
unmixed, they are, in general, a people of short
stature, small limbs and features ; where they are
mixed with English settlers, or with Lowlanders of
Scotland, the people are remarkable for fine figures,
tall stature, and great physical energy." ‡

And now as to the lessons to be drawn from all
this. We learn, in the first place, that consanguineous
unions are in all cases dangerous, and are becoming,
with our advancing civilisation, more dangerous every
year. They are therefore to be discountenanced even

in healthy families, for such unions may wake up
some pathological character which has been latent for
one or two generations.

In the next place, consanguineous marriages should
not be thought of in any family in which idiocy,
insanity, suicide, epilepsy, scrofula, phthisis, gout,
cancer, deaf-mutism, club-foot, cleft-palate, hare-lip,
rheumatism, heart-disease, chorea, neuralgia, or crime
is known to be hereditarily transmitted, or where they
have appeared in one or more generation, no matter
how far back.

And finally, we must remember the effect of "social
consanguinity," and not be too exclusive in marriage.
Let royalty renew and oxidise its blue blood to arterial
crimson at the fountain of health, even if it have to
stoop to the life-giving stream.   Let the noble improve his condition physically, mentally (and finan-
cially) by espousing the pleb of the occidental Republic.
And if the diseased *will* marry, let him be unselfish
enough to consider those to follow him : let him
mitigate his innate unfitness so far as in him lies.
Let the neurotic take unto him the level-headed, and
the feeble the robust, and brave the anger of the blind
god Cupid.

CHAPTER XX.

INSTINCTIVE CRIMINALITY.

No work, however unpretentious, purporting to treat of family degenerations, could, at the present day, be considered complete, which did not give a place to the class of instinctive criminals.

The study of this abnormal variety of the human animal, although still in its infancy, is one of the oldest of the natural sciences. Indeed, it is as old— if not even older, than man himself. Clearly the study of the criminal had its origin in the pseudo-science known to-day as Physiognomy, and this we can trace down through all races to primitive man, and not a little way further down still among the inferior animals.

The face of man, being much more expressive than that of any of the animals beneath him in the scale of development, gives a much clearer and truer reflection of the passions and desires passing in the brain behind it. Every thought passing through the human mind, cruel or kind, beautiful or brutal, is reflected more or less clearly upon the features. If a thought, passion, or desire be but transient, the moulding effect upon the features may also be transient, but if it be long continued, or if it frequently recur, be it good or evil, assuredly it will leave an indelible imprint upon the countenance that may be read of all men.

This was the teaching of the school of which the famous Lavater is the best known exponent, and in it we have to a certainty the origin of Criminal Anthropology.

For physiognomy there is much to be said. In the concrete it is largely if not absolutely true. But, lacking as it does a scientific basis, it can never be depended upon as a guide in individual cases. Nevertheless it is practised daily by every grade of humanity, from the naked savage to the ermine-clad occupant of the seat of justice, as it is also by many of the inferior animals. It is by the exercise of this instinctive power of face-reading that the dog knows when his ferocious fellow may be approached in a playful humour, and when it is wise to keep a respectful distance, or that he discovers the mood of his human master at a single look. It is by the exercise of this very same instinct that judges and juries are prejudiced for and against persons appearing before them in court, and that we ourselves discover the man whom we " wouldn't trust farther than we could throw him," without a shred of evidence against him save what we can see of the inner man shining through his tell-tale countenance.

Before man arrived at that stage of development

19

wherein he devised laws for the protection of the
feeble and good against the machinations of the strong
and evil, the exercise of this instinct was as necessary
for self-preservation in the human family as it had
been in the case of the lower animals, from which he
inherited the gift. Its presence to-day in our children
is a sad commentary on the success of our civilisation.
Had the laws devised by man given anything like
perfect protection to the just against the unjust, or
civilisation eliminated the evil instincts natural to
uncivilised man, this character must, like many another
primitive character, have grown dim, faded, and finally
disappeared as useless in the economy. But as no
civilisation can ever make the bad good, and no law
can ever fully protect the weak and honest against
the strong and dishonest—at least no law yet devised
—we find this character, which made its appearance
very early in the process of evolution, handed on from
parent to child in the highest ranks of civilised
humanity, as it was among our poor relations ages
before we came into existence.

The next step in the evolution of Criminal Anthropology, which dates from the time of Aristotle or
earlier, was attained with the establishment of Phren-
ology, that is, when man began to note how certain
instincts and passions in his fellows were associated
with certain peculiarities of cranial development. Here
the natural philosopher was on surer ground; but, as
in physiognomy, and for the same reason, viz., want
of any scientific basis, his conclusions in individual

cases could never be looked upon as accurate, however true they might be when he generalised.

The man with well-shaped skull, lofty forehead, and benevolent expression, may be and occasionally is a moral imbecile, an instinctive criminal, and the man with small brain-pan, low, retreating forehead, bull-neck, heavy jaw, and brutal expression may be and perhaps occasionally is a philanthropist; therefore the phrenologist must fail, with the physiognomist, notwithstanding the fact that if we take a hundred men of the first type there may not be an instinctive criminal amongst them, and if we take a hundred of the second type we may find but few who are not immoral, brutal, or criminal by nature.

Since the days of Aristotle, many observers have commented on the differences commonly appearing between the criminal and the average law-abiding citizen. For ages it was noticed that criminals, as a class, had characters more or less peculiar to themselves. Later it was remarked that these characters were in reality signs of degeneration, and that they were, like all other family characters, hereditarily transmitted. The soundness of these early observations has been *proved* by recent scientific investigation, and it is now recognised on all hands that the instinctive criminal is an abnormal and degenerate type of humanity.

Unfortunately, although the study of the criminal was begun so long ago, it is only in recent years that any attempt has been made to study scientifically the physical, moral, and intellectual development of this abnormal human type. In this study the Italians, with Professor Cesare Lombroso at their head, lead the van, as the representatives of the most criminal of all civilised countries properly should. Already the work has been taken up with vigour in almost every continental European state, and in America. Time was when in Bruce Thomson, Maudsley, Wilson, and Nicolson, England could boast of some workers in this field, but unfortunately at present little is being done at home to advance our knowledge of that troublesome, expensive, and interesting item of society, the criminal. It is to be hoped, however, that the recently published excellent work of Mr. Havelock Ellis,* which has attracted so much attention, will, by making known what is being done in this branch of natural science in other countries, awaken Englishmen to a sense of the great importance of the subject, and stimulate prison officials, psychologists, and others to careful research among the inmates of our prisons.

We know that when degeneration attacks the system its ravages are never confined to any one tissue or organ, and to this rule the instinctive criminal is no exception. His moral sense † is in process of

* " The Criminal." 1891.

† "Moral feeling . . . is a function of organisation, and is as essentially dependent upon the integrity of that part of the nervous system which ministers to its manifestations as any other display of mental function. . . . When it is not exercised it decays, and so leads to individual degeneration, and, through individuals, to degeneracy of race."—Maudsley in "Responsibility in Mental Disease," p. 60.

decay, but that is not his only blight. Primarily it is his moral sense which is at fault, and leads him to offend against society, but it is seldom indeed that we find any of his class at all approaching perfection, either intellectually or physically. He is a moral imbecile. He lacks the moral sense as the idiot lacks the intellectual, and in both cases the whole economy is more or less degenerate and imperfect. As we find the intellectual want in the idiot associated with physical deformity and moral perversion, and in the physically degenerate we discover mental weakness or disorder and a generally depraved condition of the whole man, so in the criminal we find the moral decay accompanied in the majority of cases by physical degenerations and deformities, together with intellectual weakness, epilepsy, and all kinds of neurotic conditions.

As Maudsley says, "The criminal class constitutes a degenerate or morbid variety of mankind, marked by peculiarly low physical and mental characteristics. . . . They are scrofulous, not seldom deformed, with badly-formed angular heads ; are stupid, sullen, sluggish, deficient in vital energy, and sometimes afflicted with epilepsy." Dr. G. Wilson stated in 1869 that "40 per cent. of all convicts are invalids, more or less ; and that percentage is largely increased by the professional thief class." * This is 10 to 20 per cent. under the estimate of the medical officers of prisons at the present day, who find less than 50 per cent. of all

* *British Medical Journal,* 1869.

prisoners fitted for hard labour. Dr. Clouston, in his
Morison Lectures delivered at the Royal College of
Physicians, Edinburgh, in November 1890, said that
he had "examined the prisoners in the Edinburgh
prison, many of whom were habitual criminals. A
large proportion of these were of the degenerate class,
mentally and bodily, and fully one half were in face,
stature, and appearance far below any minimum
standard of healthy human development." *

All who have examined the inmates of our prisons
agree that they are a degenerate and a decaying race.
They are scrofulous, and great numbers of them die
from various forms of tubercular disease. During
eight years 50 per cent. of the deaths occurring at
the Elmira Reformatory, New York, are stated by
Dr. Wey to have been due to "diseases of the chest
other than heart disease." Nervous affections also
carry off considerable numbers. Again, tissue degenerations such as are found in the gouty and
rheumatic are commonly met with even among the
comparatively young, as proved by the fact that Penta
found 44 per cent. suffering from earthy degeneration
of the tissues, and Flesch 50 per cent. with heart
disease, of which 20 per cent. actually died.

According to the report of the Medical Inspector of
English Prisons the death-rate among our convict
population is close upon a half higher than among
the general population at corresponding ages, notwithstanding that the health of prisoners is looked after

* *Lancet,* November 29, 1890

more sedulously than that of almost any other class, rich or poor.

The criminal is more nearly allied to the insane, especially the congenital insane, than to any other class, and personally he bears a strong family likeness to his near relative the idiot. In the criminal we find small, over-large, and ill-shapen heads; paralysis, squints, asymmetrical faces, deformed, shrunken, ill-developed bodies; abnormal conditions of the genital organs, large heavy jaws, outstanding ears, and a restless, animal-like, or brutal expression, all of which are common characters among the inmates of our idiot and imbecile asylums. And as there is no beauty to be found among the inmates of our asylums, neither is there any to be discovered in our prisons, which shows how well founded is that instinctive feeling of repulsion excited by the sight of the ill-favoured and deformed.

Of a truth the insane, the idiotic, the epileptic, and the criminal are bone of one bone and flesh of one flesh. They spring from like parents and succumb to like diseases. Until recently their relationship was recognised. The same devil in whose service the criminal delights "possessed" the maniac, tore the epileptic, and robbed the idiot of his God-sent reason, and the overt act of the criminal, the epileptic, and the insane alike was awarded the dungeon cell and the whip. Happily the day of corporal punishment for the maniac is past: recognising his inability to change his nature, we now humanely seclude him

from the society with which his faulty conformation
of mind unfits him to mix. How long it may be
before we come to treat the instinctive criminal in
like rational manner, it is impossible to say, but that
the day is fast approaching there is ample evidence.

The hereditary character of this peculiar degeneration, instinctive criminality, has been recognised from
very early times in the world's history. It was advocated and demonstrated by Aristotle and his follower
Galen, and it had been recognised by the Jews centuries before.

As regards mode of transmission, criminality follows
the same lines as most other family degenerations.
In some cases it is transmitted, like the suicidal
impulse, unchanged through several generations, as
was the case in the family whose genealogical tree I
here reproduce from Rossi (*Studio sopra una Centuria
di Criminali*).

R. S.

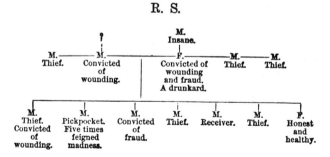

In this wretched family the insanity of the maternal
grandfather produced instinctive criminality in each of

his three children.   Where the criminal taint on the paternal side came from we do not know, but that it was strong and of some standing in the family is evidenced by the fact that every single one of the six sons of this unhappy couple proved to be instinctive criminals.   Such cases are, however, exceptional.

Sometimes, in a decaying family, in which instinctive criminality has hitherto been unknown, a generation of criminals will appear, just as a generation of deaf-mutes or epileptics occasionally appears, in families of the insane or scrofulous diathesis.   But this, again, is rare.   In the majority of cases, criminality appears in only one, two, or three members of a family, the brothers and sisters showing the taint in various ways. One will be scrofulous, or a deaf-mute, another insane, idiotic, epileptic, a suicide, a prostitute, &c., as the case may be.   Here is the genealogical tree of such a family : *

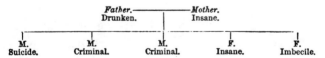

Father.		*Mother.*	
Drunken.		Insane.	
M.	M.	M.	F.	F.
Suicide.	Criminal.	Criminal.	Insane.	Imbecile.

Here we have the combined drunkenness and insanity of the parents appearing in their unfortunate offspring as suicide, criminality, insanity, and idiocy. In the following family we have the order of things reversed.   In it the criminality of the parent becomes

* *Journal of Mental Science*, Oct. 1872.   H. Maudsley, M.D.

suicide, criminality, and epileptic imbecility in his
children, and insanity in his grandchild.   This genealogical tree is from the same source as the preceding :—

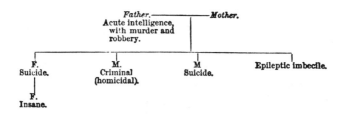

I should here remark that the prostitute ranks with
the petty criminal offender of the male sex.   She
bears all the well-marked signs of degeneration found
in male thieves, swindlers, and vagabonds, and accounts
to a certain extent for the large excess of male over
female criminals.   It is common to find in degenerate
families where the sons take to crime the daughters
take to prostitution.   In the famous " Jukes family "
of criminals, for example, the percentage of prostitution among the marriageable women down to the
sixth generation was 52.40 : the percentage of prostitution in the population generally has been estimated at 1.66.

The proportion of male to female criminals varies
considerably in different countries, but in all the
women are decidedly less criminal than the men.
Unfortunately in England there has been a marked
and steady increase in the proportion of female
criminals within recent years.   How far this may be

due to woman's adoption of the ways, manners, education, and mode of life of man is too large a question to broach here, but there can be little doubt that for the reason she is acquiring gout, general paralysis, and other diseased conditions once almost or wholly confined to man, modern woman is also acquiring criminality. If woman will invade what has hitherto been looked upon as man's sphere of labour, she must not, in the struggle for the prize, expect to escape the heat and burden of the day and its consequences.

We have already seen (p. 262) that criminals, even when we include all classes, stand midway between normal persons and the insane as regards mature parentage; and if we inquire farther we discover that age is not the only point upon which the parents of criminals resemble those of insane, epileptic, idiotic, scrofulous, and otherwise degenerate persons. Drunkenness, for instance, the most active agent of degeneration known, which we recognise as a potent cause of idiocy, epilepsy, insanity, scrofula, suicide, and every other form of constitutional degeneration, has been known since the days of Galen to be one of the most fruitful sources of instinctive criminality. " Carefully drawn statistics of the 4000 criminals who have passed through Elmira Reformatory, New York, show drunkenness clearly existing in the parents in 38.7 per cent., and probably in 11.1 per cent. more. Out of 71 criminals whose ancestry Rossi was able to trace, in 20 the father was a drunkard, in 11 the mother. Marro found that on an average 41 per cent. of the criminals he examined had a drunken parent, as against 16 per cent. of normal persons." *

Dr. Emile Laurent, in his recent valuable work †
on the inmates of the prisons of Paris, asserts that
drunkenness, alone or combined with some other
neurotic condition, is to be found in the parents of
criminals almost always, and Dr. Pauline Tarnowsky, ‡
who has made careful investigation into the mental
and physical development of the prostitute, found an
alcoholic parentage in 82.66 per cent. of 150 women
of this class in St. Petersburg whose ancestry she was
able to follow.

Of course it may be said that example and education in the home of the drunkard may account for
much of the criminality occurring in the children of
drunken parents ; but, as Prosper Lucas says, in these
heritages of crime example and education are only
secondary and auxiliary causes, and the true first cause
is hereditary influence ; as education, example, and
compulsion would fail to make a musician, an orator,
or a mathematician in default of inherited capacity, so
they would fail to make a thief.

As to the amount of insanity, phthisis, epilepsy, &c.,
met with in parents and relatives of the criminal, the
evidence is almost as overpowering as that relating to
alcoholism.   Dr. Virgilio states that 32 per cent. of
the whole criminal population of Italy have inherited

---

* Havelock Ellis in " The Criminal,"

† " Les Habitués des Prisons de Paris," 1890.

‡ " Etude Anthropométrique sur les Prostituées et les Voleuses."

§ *Op. tic.*

their criminal tendencies from their ancestors.   " Of the inmates of the Elmira Reformatory 499, or 13.7 per cent., have been of insane or epileptic heredity. Of 233 prisoners at Auburn, New York, 23.03 per cent. were clearly of neurotic (insane, epileptic, &c.) origin; in reality many more. . . . Rossi found five insane parents in 71 criminals, six insane brothers and sisters, and 14 cases of insanity among more distant relatives.   Kock found morbid inheritance in 46 per cent. of criminals.   Marro, who has examined the matter very carefully, found the proportion 77 per cent. . . . He found that an unusually large proportion of the parents had died from cerebro-spinal diseases and from phthisis.   Sichard, examining nearly 4000 German criminals in the prison of which he is director, found an insane, epileptic, suicidal, and alcoholic heredity in 36.8 per cent. incendiaries, 32.2 per cent. thieves, 28.7 per cent. sexual offenders, 23.6 per cent. sharpers." * In Russia Dr. Pauline Tarnowsky found a phthisical parentage in no less than 44 per cent. of prostitutes, while epilepsy in the parents stood at 6 per cent. †

If further evidence were wanting of the close relationship existing between the insane and the criminal, surely it is to be found in the fact that no less than 32 per cent. of all the persons convicted of wilful murder in England and Wales during the ten years 1879–1888 were found insane, and that another 32

* Havelock Ellis, *op. cit.*
† *Op. cit.*

per cent. had their sentences commuted—a large proportion on the ground of mental weakness—leaving only a third of those actually sentenced to death to undergo the last penalty of the law.

From all this it is clear, or it should be, that the instinctive criminal is as much sinned against as sinning; that, as Plato said, "the wicked are wicked because of their organisation and education, and their parents and instructors deserve punishment rather than they." It is difficult to look upon the criminal, who offends against the laws we have framed for the guidance of men and women in our social system, as the involuntary agent of a tyrannous fate. But if we will only recognise the fact that volitional power depends as much upon the organisation and healthy working of the higher nervous centres, as does the power of exercising any intellectual quality whatever, our way is clear. Once thus far, having passed the dangerous narrows of "free will," we can clearly see that the maniac who does murder at the command of a voice from heaven, his brother, who hangs himself without reasonable cause, his sister, who takes to prostitution and glories in her shame, and the other member of the family who is impelled to repeated acts of theft or violence, are equally guilty, or equally the unhappy victims of a vicious organisation. The thin end of this wedge has already been introduced, in the recognition by our courts of the dipsomaniac and the kleptomaniac, even though we are only able as yet to recognise such faulty organisation in those having good clothes and friends. But how long it will take to drive it home, and cleave the prejudice of centuries, one hardly dares to prophesy.

The records of our criminal courts prove conclusively that the various legal pains and penalties have no deterrent effect whatever upon the instinctive criminal; they have no more effect upon him than had the chain and the whip upon the ravings and violence of the maniac of fifty or a hundred years ago. Why then continue them? It would be more humane and Christian-like, having recognised the perverted moral instincts of criminals, to save them from themselves, and at the same time protect society, by secluding them at once and finally from that society with which they are organically unfitted to mix.

Dr. Maudsley says:—" It would, perhaps, in the end make little difference whether the offender were sentenced in anger, and sent to the seclusion of prison, or were sentenced more in sorrow than in anger, and consigned to the same sort of seclusion under the name of an asylum. The change would probably not lead either to an increase or to a decrease in the number of crimes committed in a year." * But on the face of it his argument is bad. In the first place, the seclusion of the asylum and that of the prison are to-day anything but the same. In the one secluded life is made as tolerable as may be, in the other it is made as intolerable as possible,

* " Responsibility in Mental Disease," p. 26.

by all means short of cruelty. In the next place, the institution of prolonged asylum seclusion, in place of repeated short imprisonments, must have an immediate and marked effect upon " the number of crimes committed in a year." And finally, the continued seclusion of the instinctive criminal would very materially limit the propagation of a most undesirable class.

Moreover, it would be much more economical to relegate the instinctive criminal, upon whom incarceration in prison cannot be expected to have either deterrent or curative effect, to the seclusion of some asylum or industrial penitentiary for good, than to be at the expense of repeated arrests, trials, conveyance to and from prisons, &c. Take the following typical cases, which I clip from the current newspapers, and let any one say whether it would not have been better in the end, both for the ratepayer and for the criminal, that on the second appearance of the criminal he had been sentenced " to be detained during Her Majesty's pleasure" in some industrial asylum.

"*County of London Sessions, March* 10, 1891. H. P., aged forty-four, was indicted before Sir Peter Edlin, Q.C., for breaking and entering a dwelling-house in Kensington, and stealing therefrom two watches, &c. The jury found the prisoner guilty. A very bad character was given to him, two previous sentences of ten years, as well as minor penalties, being proved against him. The learned chairman now sentenced him to serve a further term of ten years' penal servitude."

"*Central Criminal Court.—A Record of Crime.* A. D., aged sixty-four, pleaded guilty to stealing two pipes of the value of £3, 15s. The interest in this case centred not so much in the facts of the present offence, as in the previous career of the prisoner. Mr. Warburton said he was a notorious criminal, and, in a statement he had put in, he admitted having spent about twenty-eight years of his life in prison. Police witnesses were then called, and they stated that in June 1886 the prisoner was convicted and sentenced to eighteen years' penal servitude, but he subsequently turned Queen's evidence, and the sentence was commuted. Many other sentences of penal servitude were also proved, from which it appeared, the Recorder observed, that the prisoner had passed forty years of his life in penal servitude. His record was one list of crime, and he must inflict upon him a further sentence of five years' penal servitude." March 10, 1891.

"A curious case was heard at the Chester Quarter Sessions yesterday, when J. D. E. pleaded guilty to stealing a snuff-box and three spoons from the Grosvenor Hotel, Chester, and three handkerchiefs from a Chester outfitter. When arrested the prisoner had in his possession £95. The Recorder, Sir Horatio Lloyd, said that E. had been convicted ten times before. For the past twenty-three years he had been doing nothing but stealing and spending his time in gaol. The Recorder said he had not even the excuse that he was poor. He was sentenced to three months' hard labour." April 4, 1891.

Now, who will say that it would not have been better had these men, on their second or third offence, been sent for good to some industrial home or asylum ? Clearly their punishments had had no beneficial effect upon them, and their repeated sentences could only be looked upon as revenge taken upon them by society because of their offences. In the last of the three cases, after an experience of such punishments extending over the respectable term of twenty-three years, the criminal instinct was still so strong that, with £95 in his pocket, he could not resist the temptation to steal three spoons and three handker- chiefs ! What good we can expect to accrue from the further revengeful "three months" is beyond contemplation.

Had these three men been sent to a reformatory in their youth, and kept there, not only would society have suffered less, and the country have escaped all the expense of repeated arrests and trials, but probably a considerable number of children of the criminal class which are now with us would not have been called into being.

The following case shows how hopelessly irresistible is the instinct to crime when strongly developed :—

" At Marylebone, yesterday, William Readhead, a sharp-looking little boy, aged eleven, was charged with stealing a purse containing twelve shillings, belonging to Henry Smith, an organ tuner, of 22 Victoria Road, Kilburn. It was shown that the boy had no father, and for certain reasons he, with his brothers and sisters, had had to be taken from their mother. The prisoner had been supported by Mrs. Goschen, the wife of the Chancellor of the Exchequer, and had been in the care of a person in the village of Addington, near Croydon; but he committed so many thefts there that he had to be moved away from the neighbourhood. He was then placed in the charge of the prosecutor at Kilburn, the object being that he should attend a school in connection with St. Augustine's Church, and that the vicar, the Rev. Mr. Kirkpatrick, would have him under his special care. Since the prisoner had been at Kilburn he had committed many thefts. A purse containing twelve shillings was left on the kitchen dresser last Saturday night, and, according to an explanation the prisoner had himself given, he got up in the night, took the purse, and buried it in the garden. On the following day he was sent to church, and he then absconded, taking the purse with him. The next that was heard of him was from a telegram, saying that he had been found loitering about the village at Addington, and that some of the inhabitants had taken him in until Mr. Smith fetched him. He was afterwards brought back to London, and given into the custody of Detective Langford. The conduct of the boy had been reported to Mr. Kirkpatrick every quarter, and he was going to place the facts of the case before Mrs. Goschen shortly. Mr. de Rutzen remanded the boy to the workhouse, in order that the reverend gentleman might attend the Court."—*Times, January* 7, 1891.

In this case it was not hunger, or want, or example or trying temptation that caused the boy to steal. It was what the Americans would call pure " cussedness." In more classical, if less expressive language, thieving with him was an irresistible instinct which he was impelled to gratify on every possible opportunity. In the glimpse we have of the family we find, as we would expect, a bad heredity—immorality, and what else we know not, on both sides.

What the career before this unfortunate child may be we cannot tell. It is possible he may become a " record breaker " in the way of criminal convictions, and beat the record of the woman who was sent to prison at Liverpool recently, for the two hundred-and-eighteenth time. But of one thing we may be certain, and that is, that upon him imprisonment will have neither deterrent nor reformatory effect, while it will but poorly protect society against his anti-social instincts. His life will be one long game of hide-and-seek between himself and justice. Society will suffer much, and he will suffer more. What is needed in such cases is not punitive imprisonment, which does not improve the sufferer, and which degrades those whose disagreeable duty it is to carry it out, but lifelong seclusion in some comfortable asylum, where he may spend as happy a life as his defective organisation will permit, and which will ensure his leaving no heirs behind.

CHAPTER XXI.

SOME OF THE LESS IMPORTANT HEREDITARY AFFECTIONS.

MANY other pathological characters besides those noticed in the previous pages are transmitted hereditarily. Most of them have already been mentioned incidentally, in one or other of the foregoing chapters, but perhaps a few words might be said upon two or three of the more important of them.

ASTHMA.—This very common and distressing malady is distinctly hereditary. Observers differ as to the proportion of cases in which hereditary taint is to be found, but it may be taken that at least 50 per cent. of all cases occurring arise from hereditary predisposition.

The disorder is much more frequently met with in men than in women, about 80 per cent. of all persons attacked being males.

The asthmatic are generally thin and gaunt, with round shoulders and a peculiar circular, or barrel-shaped chest. This distorted form of chest is often inherited, and seen in infants the children of asthmatic parents, among whom it is not at all strange to meet with the spasmodic attacks of the disease itself.

Asthma is very closely allied to the gouty and the rheumatic diatheses, and not a few asthmatics inherit gout, chronic rheumatism, or heart disease, as well as the spasmodic affection. On the other hand, in gouty families it is a common occurrence for the daughters, who are not gouty, to be asthmatic, and for these to transmit gout, scrofula, or rheumatism to their offspring.

Asthma is also very closely related to the neurotic or insane diathesis, the disease being very common in families in which insanity, idiocy, infantile convulsions, chorea, and other nervous affections are to be found. In this connection it may be noticed that the father of the wretched family of suicides, epileptics, and lunatics, whose genealogical tree was given at page 18, died of asthma.

Asthmatics are themselves nervous and irritable, and generally of the melancholic temperament, as are, also, their relatives and children, among whom (as in the above-mentioned family) suicides are frequent.

Here is the genealogical tree of another family of an asthmatic father, one of whose children was a patient of mine :—

S. H.

M.
Asthmatic.

F.
Somewhat
weakminded.

| 1 | 2 | 3 | 4 | 5 | 6 | 7 | 8 | 9 | 10 | 11 | 12 | 13 | 14 |

Healthy.

Died in infancy in convulsions.

Drowned.

Epileptic.

Healthy.

Idiot.

Died in Infancy in convulsions.

Healthy.

Scrofulous.

What may have been the ancestry of this asthmatic father we do not know, but most probably it was strongly neurotic. The degenerate condition which showed in him as asthma, became fatal infantile convulsions in six of his children and epilepsy in another, while the idiocy and tubercular disease in others of his children showed the deeply degenerate character of the stock.

We are therefore justified in looking with grave suspicion upon the members of all families in which this affection occurs. Inquiry will prove that very few such families are free from gout, heart-disease, insanity, scrofula, or other sign of family degeneration.

Here is the genealogical tree of the family of an asthmatic mother. Every member of this family I knew intimately :—

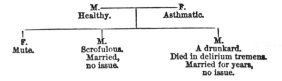

" BLEEDERS."—Of this affection I have already spoken at page 73. It is called the " hæmorrhagic diathesis," and consists of an abnormal condition of the blood-vessels, or the blood, or of both, whereby the slightest cut or scratch is followed by bleeding so profuse and uncontrollable as to frequently prove fatal. Death has followed such innocent operations as leeching, vaccination, or the extraction of a tooth; indeed, this last is the most common cause of death among " bleeders."

This abnormal state of the system is purely hereditary, and no satisfactory explanation has yet been given as to its cause or origin.   Strange to say, it is not, apparently, related to any other hereditarily transmitted abnormal condition.   Another remarkable and unexplained fact concerning this condition is, that the Jews appear to be specially liable to it.   Drs. J. Wickham Legg, Finlayson, and others have studied this diathesis most carefully, but as yet nothing has been discovered to explain why the condition should exist at all.

This hæmorrhagic diathesis is almost wholly confined to men, for the reason given at page 73, viz., that it would be impossible for a woman to survive the functions of mature womanhood in whom this abnormal condition was active.   The character is, however, constantly transmitted through the female members of " bleeder " families to their progeny, which clearly shows that although the character is not active in these women, it is present in a latent form.   This is well exemplified in the following family, whose history is recorded by Dr. Lossen and quoted by Sir W. Turner. In this family the affection is traced through three generations : not a single female member of the family was affected, yet in the second generation no less than thirteen sons of two of the females of the family were bleeders.

In this family tree the members affected are represented by capital letters, those not affected by small letters :—

THE FAMILY MANIPEL.

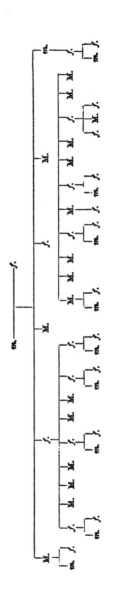

It is to be noticed, that although only one of four-teen sons of the third generation escaped, only one of nine sons in the next generation was a bleeder. This happy state of things was, in all probability, due to reversion to the healthy type taking place in conse-quence of intermarriage with healthy persons.

All that need be said of this hereditary affection is, that "women of 'bleeder' families should certainly not be permitted to marry." *

COLOUR-BLINDNESS.—This is another abnormal con-dition which is in almost all cases hereditary, yet is not, so far as has been ascertained, associated with or allied to any other hereditarily transmitted abnormal state. Like the hæmorrhagic diathesis, it is largely confined to the male sex, though not so exclusively so as that disorder. It is found in 3 to 5 per cent. of the whole male population, but in only .2 per cent., or less, of the female.

This diseased condition, as its name implies, con-sists of an inability to distinguish the various colours one from another. Red and green and other comple-mentary colours are most commonly confused. Blind-ness to all colours is rare, but in some few cases black and white only are distinguishable.

This insensibility to colour arises from an abnormal condition of the retina, and that this is a degenerate condition is, I think, proven by the fact that blindness to colours, exactly as we find it in the congenital cases at present under consideration, occurs in the earlier

* J. Wickham Legg, F.R.C.P., in "Quain's Dict. of Medicine."

stages of most cases of progressive degeneration of the optic nerve. Unlike the hereditary affection, the colour-blindness of decay of the optic nerve generally goes on to total blindness.

As in the case of the hæmorrhagic diathesis, this blindness to colours, although rarely found in women, is regularly transmitted through female members of colour-blind families to their offspring, their daughters, like themselves, generally escaping the blight. The following family history offers a good example of the usual course followed in the transmission of this abnormal condition of the visuary apparatus. It is related by Dr. Horner, who was able to follow the colour-blindness through no less than seven generations.

Those colour-blind are represented by capital letters, those not so affected by small letters :—

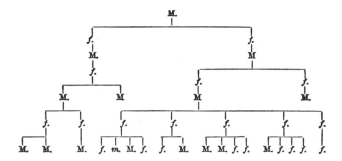

It is to be noticed, that the second, fourth, and sixth generations of this family were entirely female, and in these the abnormal character did not appear. In the seventh generation, made up of nine males and nine females, all the females escaped, while eight of the nine sons were colour-blind.

CATARACT.—This is another diseased condition of the eye which is distinctly hereditary. Like colour-blindness, it is much more frequently met with in males than in females, but, unlike colour-blindness, it is intimately related with another degenerate condition, viz., the neurotic diathesis.

This imperfection, which consists of degenerative changes in the tissues of the lens of the eye, causing opacity and consequent blindness, is in the usual course of nature a senile change : but just as we find earthy degeneration in the tissues of the youthful criminal, so in some degenerate families we find cataract in childhood, infancy, and even at birth. This degenerate condition of the eye is very common in neurotic families, and also in those of the scrofulous and rheumatic diatheses. It is a common character in the idiot, the imbecile, and the epileptic ; so common indeed is it in this class that its presence at once attracts the attention of all observant visitors to our idiot asylums. It is also common among the deaf and dumb and their relatives.

As I have said, it appears much more frequently in the male members of the family than the female, but it is regularly transmitted through the unaffected females to their offspring. This is well shown in the following family tree, in which Dr. Appenzeller follows the blight through four generations :—

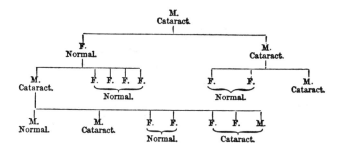

SQUINT.—This is another imperfection of the ocular apparatus which is very frequently hereditary. Like cataract, it is very common among idiots, imbeciles, and epileptics. Its presence points to nervous disorder, central or otherwise, and must be looked upon in all cases where it is not the result of local injury, not simply as a disfigurement, but as a sign of the presence of the neurotic diathesis.

The eye is more intimately connected with the brain than almost any other organ in the economy except the ear, and where disease or other imperfect conditions of the nervous centres are met with, the eye generally shows imperfect development, deformity, or degenerative change. Indeed, it is the exception rather than the rule to find perfect eyes among idiots, imbeciles, and others in whom the brain is ill-developed or diseased.

DIABETES.—This is another disease which is very commonly hereditary. Like asthma and the eye affections just noticed, it is closely allied to the nervous diathesis. It is the rule to find epileptic or

imbecile members in the family of the diabetic person.*
Diabetes is also very common in families in which
active insanity attacks some members.   Drs. Maudsley,
Clouston, and others have pointed out this relationship
between insanity and diabetes.   Dr. Savage, in a
paper which he read before the Medical Society of
London in November 1890, said that after a study of
forty patients in Bethlehem Hospital for the insane who
had diabetic relations, and ten patients who were at
once diabetic and insane, he came to the conclusion
that diabetes and insanity were closely related, and
that in such families the form of mental disorder most
common was melancholia.†

This is another of the diseases which attack the
males of affected families much more frequently than
the females, the proportion of males to females being
about three to one.   In males it generally makes its
appearance between thirty and forty years of age, but
in females much earlier, commonly between ten and
thirty years.

All families in which diabetes occurs should be
looked upon with suspicion, and should epilepsy,
idiocy, insanity, or deaf-mutism also have appeared
in the family, it is a very grave question whether
marriage should be ventured upon.

BRIGHT'S DISEASE.—This disease, which depends
upon degenerative tissue-changes in the kidneys, is
often a hereditary affection.   It is met with in various

---

* Alexander Silver, in " Quain's Dict. of Medicine."
† Society's Transactions, 1890.

forms in rheumatism, gout, syphilis, cancer, phthisis, and heart-disease, and is most certainly a constitutional disease. In some cases, as in chronic alcoholism, it appears to arise from actual irritation of the kidneys by the large quantities of alcohol they are continually called upon to cast out of the system; but even in these cases I prefer to look upon it as a constitutional affection, and class the tissue degeneration in the kidney of the drunkard in the same category with that which is found in his brain, heart, liver, blood-vessels, and other tissues of the body.

Dr. Dickinson brought a curious case of Bright's disease before the Pathological Society of London in 1889. In this case he was able to trace the disease through at least four generations. It affected fourteen out of twenty-three persons, and was traceable also in a collateral branch of the family. Not the least interesting fact in this case was that the portraits of the family, which have been preserved since the time of Edward IV., showed the pallor peculiar to persons suffering from disease of the kidney. This, however, I take to have arisen from changes in the pigments used by the artists, rather than from the existence of the disease in these early ancestors, as such a disease would assuredly exterminate a family before so many generations had come and gone.*

There are many other disease tendencies and bodily imperfections, among which might be mentioned al-

* *British Medical Journal*, May 11, 1889.

binism, heart-disease, club-foot, hare-lip, cleft-palate, stuttering, &c., which are regularly and commonly transmitted in families, but they call for no special notice here. They follow the same laws of transmission as do other hereditary characters, and the frequency with which they appear in the offspring depends upon how deeply they have been impressed upon the organism by repeated transmission, and what chance is given the *vis medicatrix naturæ* in wise marriages of leading back to the original healthy stock. Of course, in families in which any of these or other imperfections appear, the intermarriage of even very distant relatives should be prohibited.

21

INDEX.

—•—

Wey, Dr., on disease in criminals, 207 ; on causes of death in criminals, 286.

Wife-heroes, 129.

Wilson, Dr. G., on degeneracy in criminals, 285.

Wines in gout, 216.

Wisdom teeth in civilised man, 38.

Women, sacrifice themselves, 128; transmit insanity with certainty, 111, 114; are acquiring new diseases, 222; Increase of crime among, 290.

THE END.

Titles in This Series

10 Lyndsay A. Farrall. *The Origins and Growth of the English Eugenics Movement, 1865–1925*. New York, 1984.

11 Arthur E. Fink. *Causes of Crime*. Philadelphia, 1938.

12 Orson Fowler. *Hereditary Descent: Its Laws and Facts Applied to Human Improvement*. New York, 1847.

13 The Francis Galton Laboratory for National Eugenics. *Eugenics Laboratory Lecture Series*. London, 1900–1912. New York, 1984.

14 _____. *Questions of the Day and of the Fray*. London, 1910–1914. New York, 1984.

15 _____. *Selected Eugenics Laboratory Memoirs*. London, 1907–1921. New York, 1984.

16 Francis Galton. *Essays in Eugenics*. London, 1909.

17 Henry H. Goddard. *Human Efficiency and Levels of Intelligence*. Princeton, 1920.

18 Paul Kammerer. *The Inheritance of Acquired Characteristics*. New York, 1924.

19 Cesare Lombroso. *The Man of Genius*. London, 1910.

20 W. D. McKim. *Heredity and Human Progress*. New York, 1900.

21 H. J. Muller. *Out of the Night. A Biologist's View of the Future*. New York, 1935.

22 Nicholas Pastore. *The Nature-Nurture Controversy*. New York, 1949.

23 Simon Patten. *Heredity and Social Progress*. New York, 1903.

24 Hester Pendleton. *The Parent's Guide*. New York, 1871.

25 Paul Popenoe. *The Conservation of the Family*. Baltimore, 1926.

26 Robert Reid Rentoul. *Race Culture; or, Race Suicide*. London, 1906.

27 F.C.S. Schiller. *Social Decay and Eugenical Reform*. New York, 1932.

28 Lothrop Stoddard. *The Revolt Against Civilization. The Menace of the Under Man*. New York, 1923.

29 S.A.K. Strahan. *Marriage and Disease*. New York, 1892.

30 Eugene S. Talbot. *Degeneracy. Its Causes, Signs, and Results*. London, 1899.

31 William Cecil Dampier-Whetham and Catherine Wetham. *The Family and the Nation*. London, 1909.